THE ROMANTIC TOMATO

BY
Virginia B. Elliott

From Northwestern Ohio and the U-Pic Fields of Florida, laugh,
hug and taste your way through the Heart of Europe
to a charming cucina in Venice, a topless
beach in Spain, and home again.

Best Image Press & Design
Naples, Florida 34101 U.S.A.

Best Image Press & Design
P.O. Box 11983
Naples, Florida 34101
FAX (941) 514-3495
PHONE (941) 513-0606

Copyright 1995 by Virginia B. Elliott
TXU697-326
Publication Date January 1997

ISBN 1-883782-02-3: $9.95

Books by Virginia Elliott

3 Tomato 4, Best Image Press & Design

How to Board Up Your Kitchen, First Ed.,
Best Image Press & Design

How to Board Up Your Kitchen and Cook From a
Hammock, Top of the Mountain Publishing

The Romantic Tomato, Best Image Press & Design

The author's articles, essays, poetry and fiction have
appeared in regional and national periodicals

Table of Contents

Are you kidding?!

We managed to make a perfectly workable index, with identifiable titles for recipes. You and I, I am sure, cook by the seat of our pants, and shop according to market specials, our current budget, and the mood we are in at the moment, right?

Get in a romantic mood by scanning the index for romantic cities and then sitting down and enjoying my memories.

Cuddle with someone while you're at it.

Virginia B. Elliott

Angels at my fingertips!

Without the constant encouragement of friends who tasted 'til their taste buds rebelled; my son David, who can spot a run-on sentence in a dumpster of rough copy, and who crammed a modicum of computerese into my stone age brain; my long-suffering husband who finally (almost) got sick of tomatoes, late dinners and carry-out foods, *The Romantic Tomato* would have been blocked forever in our nostalgic memories.

Humble and hearty thanks to Mary, a creative cook who often stretched my recipes to a higher level; and who loaned me her husband Grant for hours of U-Pic field harvesting; and to Grant, who started the whole idea when he deposited about a bushel of ripe tomatoes on my doorstep one weekend!

Loving appreciation to Barbara and Lillian, pals who often kept the cash box straight at signings for my first books, and who promoted them to anyone who would stand still for five minutes.

Special thanks to independent shop owners and booksellers all over Naples, Florida who took a chance on a novice by graciously hosting appearences for me.

I am truly grateful to the many *Books A Million* and *Barnes and Noble* store managers and coordinators who also promoted my books and included me in many special events.

What a support system!!!

Virginia Elliott often traded her party recipes with friends and her husband's clients. But when the requests became a burden, she wrote the modest *3 Tomato 4* booklet and had it produced by the little printer around the corner. When the third printing sold out a writer chum said, "Why don't you stop this 'print shop' caper and make a real book?"

Not knowing the first thing about 'making a real book', she just up and did it. So a ninety page book, ***How To Board Up Your Kitchen and Cook From a Hammock*** was born and sold out 1500 copies in seven months in Naples, Florida alone. Then a major Florida publisher urged her to expand it to 256 pages and let him produce it. This was so successful that she decided to expand the little tomato booklet and here 'tis!

The recipes are steeped in Midwestern family life as well as humorous and romantic experiences while touring Italy, France, Spain, and Austria.

Recipe packets for the recommended anti-carcinogen foods, and her clever patterns for using scrap lumber and fabric to make living easier and less costly are available by mail order (see the order blanks in the back of this book).

Another book on the horizon? Absolutely!!!

TOMATO LORE
Historically Speaking

This book has been inspired by my Grandparents, Anna Mary and Wallace Spitler, who reared me in value packed, Northwestern Ohio during the Great Depression. They would be horrified at the plowing under of acres of gorgeous ripe and green tomatoes at the end of the tomato harvesting seasons in South Florida. Such a waste! When I inquired, a local grower said it was cheaper to plow under than to pay hands to pick after the vegetables had become too mature to transport to northern markets; also, that insurance costs were too high to open the fields to indigent, hungry people for free picking. When we pick, we share with those who need the food but cannot get to the fields. Try it, it'll make you feel wonderful!

TOMATOES are botanically a berry (fruit) ...first found in the South American Andes...named by Aztec and Mayan Indians of Mexico...brought to America by Spanish Conquistadors in the 16th century. Early Americans considered them poisonous, and when they did eat them, used sugar instead of salt to enhance. There are now over 1,000 varieties in all shapes, sizes and colors. Over 35 million American families grow them in their door yard. Many retirees and city folk grow them in half barrels on high rise condominium balconies. French chefs called them "Pomme d'amour" or love apples. So you see, the *Romantic Tomato* is not such a wild idea after all.

1

BASIC RULES

KEEPING FRESH - NEVER, NEVER, REFRIGERATE even the ripe ones. Greens will ripen to perfection on your counter *if* they have never been chilled. Hasten ripening by laying a red apple or ripe banana in a bowl with green tomatoes. To serve chilled you may refrigerate enough for each meal for 2 hours before serving. As for me, I love a field fresh, sun warmed aroma and taste, as if straight from the garden.

TRANSPORTING: Do not include the stem when you harvest tomatoes; it will puncture other tomatoes and the wounded one will rot quickly (Remember the adage, about one rotten apple in the barrel). Take a firm but gentle hold of the fruit in your whole hand and twist gently so as to take the fruit from the stem, and not the stem from the plant.

My friend Grant, who harvests the Florida U-PIC-FIELDS throughout the growing seasons, used to carry boxes of tomatoes back north to friends. He taught me to layer pink ones separated by flattened grocery bags to transport north. They will ripen nicely.

Grant would pack green tomatoes separately for frying, pickling, pies or preserves. They should be processed as green within two to three days after picking, or they will start to ripen.

Incidentally, if an early freeze threatened, I remember my grandparents pulling up plants, roots and all, with pinks and greens still attached, and laying these plants, over rafters in the attic. They would ripen and we would have tomatoes to harvest daily for several more weeks.

TOMATOES GALORE!!!
Pick, Cook, Freeze and Store

What a great opportunity for tomato lovers when local growers all over the country open their fields to U-Pickers at a generous mark down per pound. Even if your own garden is overflowing don't let one tomato go to waste! Invite a friend to share the sunshine and some good old fashioned talk. Get double value out of the activity; straighten and bend as you go, keeping your knees flexed, much easier on your back and great for your waist line! Pick in the cool of the morning.

Take several spare grocery bags, or some sturdy boxes from your grocer, and pick about 20 pounds of ripe, crimson fruit. Pick about 10 pounds of green ones too, for processing as well as to let ripen for table use. Remember about those stems!

Pick the really ripe ones in one bag, (not too many that they squish), pale pink in another and green ones in a third bag (or use boxes). Back home, get ready to process your bounty. You should have been collecting pint or half-pint plastic containers as they freeze fast, are easy to stack and store in the freezer and thaw quickly when you are ready to create your famous spaghetti sauce or chili. But you don't need a jillion of them. Use five or six to freeze on your freezer shelf and transfer to Seal-a-Meal bags. I use one-pound and half-pound margarine tubs because sometimes a recipe will only call for a scant cup of sauce. They freeze and thaw quickly. If you have a Seal-a-Meal contraption you can save freezer space by using small portion bags. They will freeze quickly and store flat or standing up.

ABOUT THIS BOOK'S FORMAT:

Most cookbooks offer orderly recipe sections set forth in a Table of Contents, starting with appetizers, soups, snacks, sauces, entrees, etc., etc. Phooey! When I plan a meal, the first thing I think about is, what do I have to use up, or what's economical in the market and then I plan an entree on that criteria, and build the meal around it the same way. So the index will get you to the item you want, even though it might be an appetizer near a soup. We will start with Basic Sauces which will become whatever you have in mind and go on from there. The index will lead you to canapes, soups, sauces, entrees, casseroles, ethnic dishes, sweets, sours and sandwiches. Enjoy your crop! The tomatoes are waiting!

5

> Tip: Unless otherwise directed,
> "dice" means about 1/4 inch.

BASIC TOMATO SAUCE
Makes about 10 pints

10 pounds red ripe tomatoes.
1 large bunch celery (6 cups diced). Cut off about three inches of the bottom end and the leafy end to save for salads and braising for use as a side dish.
3 or 4 green bell peppers (4 cups, diced).
10 medium sized onions (6 cups, diced).
6 Tbsp white sugar, to cut acidity.
1/8 cup salt. (Omit or cut, if you're watching blood pressure in the family.)
1 Tbsp white pepper, or less-your taste.
1 cup fresh basil - OR 1 Tbsp dry.

To peel tomatoes easily, scrape them with the back of a table knife, or plunge into boiling water for just a minute on the end of a long cooking fork. If you have a microwave, lay a layer of tomatoes on a glass platter, process at full power for 1 minute and let set two minutes. Skins will slip easily.

Dice celery, onions and green peppers into a huge bowl, or clean plastic dish pan, whatever you have that will hold the whole recipe for mixing.

Now, peel the tomatoes. If they are real seedy, give them a poke with the tip of your knife and dump the

6

seeds. Then dig out the stem end of the core. Cut them so your chunks are about an inch cube and add them to the bowl of vegetables along with the seasonings. With two spoons or your two clean hands – my preference – plunge in and mix well.

NEXT YOU COOK: Microwaving

Use a huge four quart Pyrex® or microwave safe bowl and fill it to within 3 inches of the top. You will have to process in about three batches. Cover with wax paper. You will have to change to fresh paper halfway through when you stir. Process 8 minutes on high. Stir. Change paper if necessary. Process eight minutes more on high. Stir. Process eight to ten minutes on 50% power; sauce should simmer. Taste, and if not cooked as done as you like, try another 5 minutes on *Simmer*. The batch should have just come to a nice gentle bubble.

Microwave cooking continues during a 5 to 10 minute "rest" after cooking in this large a batch of food, which will complete the cooking. Even in sauces, we do not like our vegetables cooked to a mush; the times given here should be about right. Remember, in creating your favorite recipes from the Basic Sauce, it will cook more.

Stove Top Cooking

Never, never, use aluminum or black iron for acid fruits such as tomatoes. Use Corning Ware®, Pyrex® glass, porcelain or stainless steel – as large a pot as you can beg, borrow, or buy. Of course, you can split the batch and use smaller pots. Bring to a boil for about 4 minutes. Turn heat to *Simmer* and, setting pot off heat, skim. Put back on heat, cover and simmer for ten minutes. Remove lid, stir and skim and continue to simmer *uncovered* for about 20 minutes to half an hour. Then cool for half an hour, stirring often to help it to cool.

NOTE: Stove top times will result in a more *cooked* sauce, but you need the extreme heat to kill bacteria. Ladle into freezer containers and freeze.

Now that you have 10 pints - more or less - of Basic Sauce in your freezer, here are many recipes you can make as given or *you* can "springboard" to innovative combinations of your own, as well as using in your own old family recipes or new ones from a friend.

> **TIP:** Remember you can use a pint of basic sauce in any recipe calling for a #10 can of stewed, sliced, or diced tomatoes.

FROZEN WHOLE TOMATOES FOR WHEN THEY ARE OUT OF SEASON

Raw tomatoes do not freeze well, as they lose taste and quality. However they can be frozen and stored in tomato juice as follows:

6 small tomatoes will pack nicely into a 1 pint flat, square freezer container. For each 6 tomatoes peeled and cored, add enough canned tomato juice to fill the container. Leave 1/4 inch for expansion. Using these measurements, bring juice to a boil and then ladle back into container. Seal and freeze. Use in any recipe calling for canned whole tomatoes.

FROZEN TOMATO SLICES: Use these in cooking within 6 weeks, and they do better if lightly salted and sugared before freezing.
Peel red ripe tomatoes, slice a scant half inch thick. Freeze in plastic bags, in single layers. Use to decorate meat loaves, roasts, or casseroles. Also use in omelets and quiche when tomatoes are off season in the market.

NOTE: All recipes are easily adaptable for *vegetarian diets,* by substituting sauteed cubes of Tofu, or vegetarian meat substitute from health stores, or simply adding grains and/or dried beans.

Where salad oils are designated, CANOLA OIL AND VIRGIN OLIVE OIL ARE RECOMMENDED BY GASTROENTEROLOGISTS. Just remember all oils contain fat.

VARIATIONS OF BASIC SAUCES

FIRST OFF, Be sure to notice that the basic sauce ingredients and rules makes just that, a basic sauce. Then you can make this sauce fit any ethnic recipe by simply adding the seasoning ingredients that are especially germane to that country's cooking. For instance, *cominos seeds and chili powder for most Mexican dishes; garlic, oregano and the corianders of Italian cookery; fennel, and feta cheeses for many Greek dishes; ginger, soya, and five spice powder for Oriental dishes. Got the idea?*

9

Now, for some variations. Is your family crazy for Salsa? Try mine. Glass jars of the many different grocery store versions are very good, but I have never been able to open a jar of anything in my life without embellishing it. The trick is, whatever you add should be diced as itzy bitzy as you can manage. It can be anything that grows and can be eaten raw, from apples or avocado to zucchini. Yes I said apples, very solid, tart ones, like Granny Smith®. Spray them with a mist of lemon juice to keep them white and add a dash of crazy salt seasonings and toss together before adding them to Salsa. It is a tantalizing taste sensation!

FOR MY OWN SALSA: I thaw a half pound margarine tub of my basic sauce and start chopping and adding whatever I have on hand in the crisper of my fridge.

For instance: *Well-scrubbed cucumbers*, scored with a fork, not peeled; chopped and tossed with a bit of fresh chopped dill. *Fresh cilantro and tender hearts of celery* and a tablespoon of whole cominos seeds (Cumin). Add an extra *tomato* or two, seeds and juice poked out and pulp chopped small.

A fresh avocado, not too ripe, chopped small. Since avocado is so bland, it sometimes will help to mask the bite of an extra hot salsa.

If you find salsa, yours or store bought, is a mite too juicy for easy dipping, add a tablespoon of *canned tomato paste*. Or stir in two or three tablespoons of seasoned bread crumbs. Store the leftover tomato paste in a sandwich baggie and put in freezer, ready for the next time you only need a spoonful. Experiment! Be daring. Tease palates.

10

And get this, Salsa isn't just for drowning tortilla chips in. Try heating it, and tossing it with a pound of freshly cooked pasta. Top with grated mozzarella instead of parmesan. Wow, that's wonderful!

The French, Italians, and Californians all have a way with sauces. Here is a simple farm sauce for over omelets, seafood, fowl, veal or frog legs. In French, "Sauce Provencal," but me, even though one of my great grandmothers came from France, I'm mid-America Country (Missouri and Northwestern Ohio)!!

SIMPLE SMALL TOWN SAUCE FOR MEAT

1 pint of Basic Tomato Sauce, thawed.
2 buds crushed garlic.
1 chicken bouillon cube, or 1 Tbsp bouillon granules.
1 cup dry white wine.

Simmer and stir together rather briskly 'till it reduces to about 1/3. Careful! Don't scorch. Dribble this reduced sauce over each meat serving; top with minced parsley.

BONUS: If you have sauce left over, freeze it in a small container or a sandwich bag. It is wonderful over burgers or baked potato for a quick lunch.

Vegetarians who love pasta can start with the following simple recipe and then add any variety of veggies that are fresh in the market or garden. In a pinch canned or frozen will also work. Be sure to freeze juice from canned for future soups.

11

MARINARA SAUCE
A simple meatless Italian sauce

1 pint basic sauce, thawed.
5 peeled garlic buds, more or less to suit
taste. We love the stuff!
1 cup fresh, chopped parsley
1 medium size carrot, sliced thinly, optional.
1 6 oz. can tomato paste

NOTE: I cook the carrots in microwave first. They sweeten and cut the tart of tomato, without sugar.

Puree first 4 ingredients in blender and put in sauce pan. Then add the 6 ounce can tomato paste.

Simmer about 30 to 40 minutes. Stir often. If getting too thick, add bouillon bit by bit. Use a stingy hand to add 1 or 2 of your favorite herbs, just a pinch each rosemary, oregano, sage, thyme, or a couple of bay leaves. *Not all!* Adjust salt and pepper, add 1 Tbsp chopped fresh basil. (Double pinch of dried will do if no fresh handy.) Simmer another 10 to 15 minutes

MARINARA WITH MUSHROOMS

To the above recipe, during the last ten minutes of cooking, add as many mushrooms as your family likes. We add a pound box, but you can also use the canned stems and pieces if you wish, but be sure to drain off water into a half pound margarine tub & freeze to add to a soup in your future! Incidentally, allow me to digress a moment here. You know those tiny little glass pots in which you buy grated ginger, minced garlic in oil and/or

Well I keep these empty jars on hand, ready for itty bitty amounts of juice and seasonings I cannot bear to discard. Keep them together by stowing them in little 4" by 6" plastic baskets. Leftover cheeses store here too - in freezer or fridge. If freezing, be sure to leave about 1/8 inch of headroom so jars won't burst as they expand in the freezer.

GROUND MEAT SAUCE

Brown 1 pound of your choice of ground meat. In a colander, drain browned meat and rinse with a cup of hot water, fat will go down the drain! Proceed with marinara recipe above. Add mushrooms if you like.

OR, choose one of these variations: use your favorite meatball recipe, (brown them first), use browned strips of chicken breast, or a handful of shrimp; whatever!

SAUSAGE SAUCE

Great for Lasagna: Cut and brown bite-size chunks of Italian Sausage, hot, or sweet or some of each, suit your family tastes. One and a half pounds total. Then proceed with the Marinara Recipe. You will not only want to drain all fat off before adding sauce ingredients, but skim fat during cooking as sausage will keep giving off more fat as it cooks in the sauce.

I find that sausage will cut easily into chunks if I let it almost freeze first. **Also,** When making Lasagna for a mob, I spread the sausage chunks out on a cookie sheet and put them in a 400° oven and watch carefully. You can brown 3 or 4 pounds in a hurry this way. Of course this will call for a bunch more basic sauce.

13

Another way to reduce fats in ethnic sausages is to boil them whole in beer first, then saute or broil. The same oven procedure cooks up a zillion meatballs in a hurry too. And of course you do know about getting a head start, don't you? Make the meat sauce the day before and chill it so the fat all comes to the top and can be lifted off and discarded easily.

You can also add mushrooms and ground beef to any basic marinara sauce, mine, yours or store bought. If you start with *2 or 3 pints of sauce* it will feed an army! The finished meat sauce will also freeze well, if you should have any left over. Bet you won't!

HERE IS AN IMPORTANT CAUTION: You should be aware that sausage meats, because of their high fat content, will not remain fresh tasting when frozen if not used in ten days to three weeks at the most, even in a sauce. I learned this the hard way. Preparing for a scheduled writer's party I made my sausage sauce a month ahead of time and the rancid odor that filled the house when I started to heat that sauce was really awful! Thank God for my neighborhood Pizza Parlor delivery service. But my frugal country soul nearly died when I had to pitch that gallon of sausage sauce!

> **TIP:** A measuring cup of water and a teaspoon of savory pie spices boiling away on a back burner will always clear the air of any unpleasant cooking odors.

HAMBURGERS AND SIMPLE BASIC SAUCE

A bunch of us were talking about simple favorites from our childhood. One of mine was my Gram's thick hamburger patties, seared quickly in a hot skillet to seal in the juices, simply smothered with lots of sliced onions and her home canned tomatoes. Use your basic sauce here, simmer away for about ten minutes and then just thicken it with a tablespoon of corn starch in about three tablespoons of *cold* water to make a wonderful tomato gravy for over the patties and mashed potatoes. Season to suit yourself. This is rib stickin' country stuff, and once in a while this country cookin' is just what a stressful day calls for. Cooking always seemed to lower my Gram's dander when ever it got up!

Sometimes what starts out as cocktail snacks can turn into an unforgettable meal. And meatballs from every culture in the world can double as a starter, a sandwich or a meal. Read on, after the following recipe.

BLACK BEAN MEATBALLS

1/2 cup salsa (mild or hot to suit your family taste) from deli, ethnic food aisle, or recipe in this book under "Sauces"
1/2 cup Pepperidge Farms® corn bread stuffing mix.
1 can black beans, well drained.
1 lb very lean ground beef.
1/2 cup finely diced onions.
1/2 cup Pepitas (salted pumpkin seeds)

15

In a large mixing bowl pour the salsa over the stuffing mix and allow to stand about fifteen minutes to soften the bread crumbs.

Add rest of ingredients and mix and mash thoroughly, with a strong "squeeze" so the beans really get mashed up. Form into one inch meatballs.

At this point, if you have *not* added ground pork or sausage on your own you may freeze the raw *beef* meat balls for up to six weeks. Then bake on cookie sheet (40 minutes at 375° or until thoroughly cooked, *not rare,* for serving as a canape. If made with sausage, do not freeze.

OR - you may bake on cookie sheet at once to be ready to serve when guests arrive. **OR** you may bake as above and then freeze until day of party and thaw in microwave - 6 minutes on thaw, and reheat in microwave, 3 minutes on high, or until piping hot.

When you live in Florida, everybody you know from above the Mason-Dixon line wants to visit you. And if you are an Elliott, you always say, "If you are ever in our area, call us and plan to visit." You have to be prepared!

All of the above ingredients were on my pantry shelf and the meat was in the freezer when friends whom we had met on a tour to Europe announced their presence in the area at 5 p.m. We, of course, said, "Come on to our house, now!"

I began to thaw the frozen ground meat in the microwave and make up the meatballs. We figured we would have a quick appetizer and some cool libation and then go out to dinner. But our friends had left frigid Maine three days before and once over our state line could not resist a stop at one of our wonderful Florida produce stands. They came bearing gifts of several kinds of lettuce, avocados, tomatoes, huge Florida sweet onions, luscious strawberries and a monster hybrid mango, and they were drooling! We ended up making a huge salad, heating a loaf of garlic bread from the freezer and, using the meatballs as a sort of mini entree. We dined leisurely, shoes off, feet propped on cushions, around the coffee table in the family room, ending up with strawberries and chunks of juicy mango dipped in cinnamon sugar. What a wonderful evening it was, reviewing our European tour and getting better acquainted. No fancy table setting to impress, no dish washing to speak of. I used paper dessert plates left over from a party, for the meatballs, but I did add a touch of class. I stuck the salad plates in the freezer to chill while we all rinsed and cut up the salad stuff. Our male guest prided himself on his own salad dressings so he raided my condiment shelf and proceeded to "create" as we talked.

We piled the prime strawberries in a faux milk glass compote from the florist (I save everything), and set the compote on a piece of dark green pottery from a flea market with the mango slices arranged around the foot of the compote. Everybody had a tiny cordial glass filled with the cinnamon sugar for dipping.

Since we are talking about ways to start a meal, or turn a starter into a meal, how about soup?

Bear with me now, as far as formatting this book is concerned. The index will take care of any resulting confusion, but I have scattered soups all through the book, because they can be enjoyed as a starter, a luncheon, a hearty supper or even breakfast. The following three are hearty enough to stand alone - alongside a basket of bread.

BORSCHT

This hearty soup should be made the day before
serving for two reasons: the flavors will improve and you
can skim off the fat easily.

3 lbs lean beef, Chuck, bottom round or shanks.
Leave the meat in one piece. If you want to reduce
fat, trim well and dig out and discard the marrow
from the shank bones. We love the hearty full beef
flavor of shanks.
1 small head of cabbage, shredded (3 cups).
2 large onions, sliced thinly
2 or 3 sticks of celery, finely sliced on the slant:
use some celery leaves if you have them.
5 or 6 fresh beets with tops.
2 cups canned tomatoes or fresh ones, diced
1 tsp Nature's Own® Seasoning or Jane's®
Mixed up salt.
Pepper to taste, try 1 tsp.
1 level Tbsp sugar.

In a large heavy pot, put 1 Tbsp cooking oil to
brown the meat on all sides, together with the onions. If
watching fat, use non-stick spray.

In another pot boil the beets and greens. Peel and
shred beets, chop greens fine. Strain beet liquor through a
cloth or a coffee filter in case a bit of sand stuck to the
greens. *See note below for a speedy version.

Smash up tomatoes. Combine everything,
including beet greens and liquor and enough more water
to cover all by an inch. Barely bring to a boil, then simmer
three hours, covered with lid. Stir now and then.

Refrigerate overnight and remove solid fat. Reheat slowly. Slice or chunk up meat onto a hot platter. To serve, place a couple slices of meat in soup plate covered with heated soup and a dollop of sour cream.

Serve heavy black bread with this and rent a video of The Ballet Russe for after dinner.

*EASY, FAST BORSCHT: Use 1 can of shredded beets, including juice, instead of fresh ones. Add about 6 outer dark green leaves of romaine, shredded, for the last 30 minutes of heating.

ROOT SOUPS

When traveling in central Europe, we found a much wider use of all root vegetables than city folks here in the States seem to use. It reminded us of the winter root "mound" of our childhood. In late fall, before the ground froze, Grandpa would dig out a sort of bowl in the field nearest the house, about a foot deep and five or six feet long by 2 or 3 feet wide, depending on how much of a root crop he had to preserve for the winter.

Rutabaga, white turnips, parsnips, carrots, and cabbages were then laid out in this bowl and covered over with soil to make a mound. Even in the frozen dead of winter, he could go out there and chop away and bring in perfectly preserved vegetables. Of course he had to remember where he had stashed the various kinds so as not to have to dig all day to find the carrots. But he never failed to bring in what ever Grandma asked for.

This was a pioneer forerunner of today's freezers, except that today we wash and blanch and chill and package for the freezer. Seems like Grandma's way was easier, she got Grandpa to do the hard part. And yet, our

19

way cuts down on the last minute cooking chores just before mealtime. So much for the "good old days!"

HAM WITH ROOT VEGETABLES

This hearty meal in a pot can be served any time of a cold winter day, (even breakfast) and the older it gets the better it gets.

Start with a couple of ham hocks, a leftover shank from a baked ham, or just some cubed ham meat, what ever you have on hand - use it.

Put ham in a pot with about a quart of Basic Tomato Sauce, or canned tomatoes.

Add 1/2 cup each diced onions and celery.

NOTE: If celery is from out of state and expensive, just use more onion and 1/8 teaspoon celery salt.

Bring to a boil, turn down heat and add the following:

1 cup each diced parsnips, white turnip, yellow turnip, (rutabaga), carrots and potatoes. Cook at a simmer until all vegetables are tender. Adjust seasonings, being careful of salt, as the ham has lots. We sometimes add some chopped raw cabbage to this. It is a hearty soup and a touch of nutmeg won't hurt it a bit. My husband likes black bread and cold beer with this. It is a wonderful cold weather meal with a hot apple pie for dessert. If you are too busy to bake, bring one home from the store, sprinkle just a tiny bit of sugar and cinnamon over top and heat it in a warm oven during dinner. Put some frozen vanilla yogurt on each slice and send the kids to their room early 'cause their Dad will be making sexy suggestions before you know it.

Another way to use up leftover ham or roast pork is in a lima bean soup. We sometimes forget how delicious lima beans are and they are loaded with nutrients. My family likes the little green limas, but I love the big old fat butter beans.

QUICK AS A FLASH LIMA SOUP

Start with whatever amount of leftover ham or roast pork you have on hand....oh, maybe about 2 cups of chopped meat. If you have roast fresh pork, add a dash of Wright's® Liquid Smoke or a quick shake of Smoke Salt seasoning, it'll taste like ham!

Add meat to a pint of basic sauce (or canned Tomatoes)
Add two 10 ounce boxes frozen Fordhook lima beans (or butter beans if you invite me.)
Add 1 envelope French Onion Soup mix.
Add the amount of water called for on the Onion Soup Mix packet.

Now bring it to a gentle boil and then simmer about 20 minutes. We like to dump a handful of seasoned croutons into our soup bowls instead of crackers. And sometimes, when I feel like taking the trouble, I thicken the broth at the last minute with 2 eggs whipped up with 1/4 cup cold water and then 1/4 cup of the broth and then stirred into the simmering soup...sort of an egg drop broth. And I have also added, at the last minute, some shredded romaine leaves or spinach leaves. Try it. One of my kids did not like the "mushy/starchy" texture of butter lima beans but ate the soup happily when I added a can of sliced Chinese water chestnuts so there was some crunch.

21

You know we do get in ruts of our own making from time to time. I find myself thinking only Mexican or Cuban food when I pause at the kidney and black bean shelves in the market. But one day my hubby wanted some bean soup for lunch and I had only 2 cans of red kidney beans in the pantry. Well, nothing ventured nothing learned, right?

KIDNEY BEAN SOUP

2 cups slivered onions - thin strips.
1 cup slivered green peppers
2 hotdogs, leftover or brand new, sliced paper thin. Or one lonely kielbasa, use it up!
1 can beef broth or 1 1/2 cups bouillon from cubes. 1 pint of Basic Sauce or canned tomatoes
2 cans kidney beans

Saute the onions, green peppers, hotdog slices in a non-stick pot, just until things start to brown. Add rest of ingredients and simmer about 20 minutes, covered. Season to suit your family. We like 1 tsp paprika, 1 tsp garlic powder (or crush 2 buds fresh garlic into the saute) and enough pepper sauce to make your nostrils clear. Suit yourself.

And I'll tell you something else about this soup. Try serving tortilla chips on the side to crumble into it instead of crackers. Buy the baked kind of tostadas without the fried fats.

If you add handful of leftover pasta and 3 or 4 crushed garlic buds, your Italian neighbors will love it. Or try mixing cultures as Marco Polo did when he brought pasta to Italy. Read on.

Our whole family loves Chinese foods and especially Won Ton soup. Now I reckon your neighborhood Chinese chef would tell you to get real if you tell him about this, but do try it anyway, the pros don't always know it all.

WON TON SOUP WITH AN ITALIAN ACCENT

Before you wonder if I left out an ingredient, this soup does *not* use tomatoes in any shape or form. I just want you to know how flexible I am. So there. Besides it's so good I had to share.

> *1 quart of chicken broth, from your freezer, cans in your pantry, or made up with bouillon cubes.*
> *5 whole raw lasagna noodles, broken up into roughly 2 inch crazy shaped pieces.*
> *OR, a handful of frozen raviolis or tortellinis.*
> *2 cups chopped fresh spinach leaves*
> *OR romaine, or dark outer lettuce leaves.*
> *1 tsp lite Soy Sauce.*

Bring broth to boil, add noodle pieces and cook until al dente. Add spinach or romaine and soy sauce. Cook 5 minutes more. Serve with a good pot of tea.

Normally Italians use tortellini as an entree and along about page 74 in this book I suggested a way to serve it as a canape, but try the following soup idea too. Very filling, really delicious and just about as quick and easy as pouring a glass of wine and opening a can. Incidentally, try a handful of tortellini added to any soup you want to extend. It will not only increase your servings but add a touch of classy flavor.

TORTELLINI SOUP

5 buds of garlic, minced
1 Tbsp Olive oil.
1 quart of Basic Tomato Sauce
2 14 oz. cans beef broth
1 can lentils
1 box (2 cups) dried tortellini (cheese or meat,
doesn't matter.)
1 glass of wine, (in the pot or you, maybe both).

Saute garlic in oil. Add basic sauce, beef broth and lentils with their liquor. Bring to boil. Add tortellini and cook until al dente. Have hot bread ready to dunk.

BISQUES are a delicate way to treat soup when you want to be stylish. A bisque is a simple soup gussied up with cream, and maybe a handful of chives scattered over, or a slice of cucumber floating on top, or a blush of grated nutmeg: and hang the calories, use real cream, **or save your arteries,** do it my way, as follows:

SHRIMP BISQUE

1/2 pound fresh shrimp, cleaned and diced.
OR 1/2 pound ready cooked shrimp, from freezer
section of market.
1 Tbsp good butter flavor margarine
1 pint of Basic Tomato Sauce
1 pint of chicken broth
1 can Cream of Tomato Soup
1/2 cup sherry
1 cup no fat sour cream.

Saute diced shrimp in margarine in a pot that will hold all ingredients.

Add tomato sauce, broth, tomato soup (do not dilute soup) and sherry. Simmer and keep stirring gently to blend well. At the last minute before serving, whisk in the sour cream (Here comes my husband with the nutmeg shaker again.).

DOUG'S SWEET POTATO BISQUE

Bake a couple extra sweet potatoes and make soup the next day.

2 cups mashed sweet potatoes
1 can cream of chicken soup
1 soup can skim milk
1 1/2 cups chicken broth, defatted.
1 can Tomato soup plus half a can of skim milk
3 Tbsp melted margarine (butter is delish, but will your diet permit it?)

Low fat tip: Mix 3 tablespoons butter flake flavoring with 1 tablespoon canola oil and use instead of margarine or butter for taste without fat.

We like to use some white pepper in this, but go lightly, it seems to be a mite peppier than black. We add just a dash or two of celery salt or Aunt Jane's® Krazy salt; not both and not a lot, as the soups contain salt. Mix everything up in a sauce pan over medium heat and whisk alot so it gets smooth and doesn't stick. At the last minute whisk in 1/2 teaspoon nutmeg. We like toasted seven grain bread with this; you know, the kind with whole grains and nut meats so it is crunchy.

So you are saying, "This is supposed to be a tomato book." No actual tomatoes in sweet potato soup, right? Well, serve sliced ripe tomatoes instead of a salad. What could be easier? Those tangy tomatoes compliment the creamy soup perfectly. So there.

Did you ever hear of *baked soup?* On one of our band trips to Italy, our driver got lost and we missed lunch altogether. We were scheduled for an early gig in Bari. After checking into our hotel the decision was made to try to find a hearty enough late lunch to tide us over instead of dinner. Just around the corner was a small restaurant where the proprietor was just closing up for siesta.

However, we prevailed on him to feed us. He quickly carried a huge heavy black iron roaster or casserole to a table next to us; set out a tray of soup bowls and a ladle. I had seen him pull the pot from the oven, but this was soup, nonetheless. I tasted the recipe out of that heavenly concoction and we have all called it *"baked soup"* ever since. He gave us a huge metal tray of crispy bread chunks from the oven too. It was a feast, I tell you.

His English was worse than our Italian so we never did find out if he baked the soup from scratch, or if the oven was his way of keeping it over until the evening dinner hour. We ordered a bottle of wine and devoured the soup which we ladled up ourselves, while our host snoozed in the corner. Nothing interferes with siesta in Italy, Spain and Portugal. And it's a good idea anywhere, even Ohio. We all learned a hot tip that afternoon. When reheating soup, you can avoid constant stirring by using this oven method. About an hour at 250°.

Several of us from that trip have agreed that starting a hearty soup or stew in an ovenproof pot on top of the stove and allowing it to finish in a low oven saves constant stirring and gives us a free afternoon for hugging our kids or reading a book before dinner.

BAKED SOUP FROM ITALY

6 quarts of water.
1 lb of ziti (Large macaroni).
1 lb hot or mild, (your choice) Italian sausage.
1 lb lean ground beef.
As much fresh garlic as your family likes. We use 6 to 8 large buds, grated into the ground beef when browning. But do not ever scorch garlic! Yukky!
3 large onions, sliced
6 fresh tomatoes, peeled and quartered. Or a pint of Basic Tomato Sauce, Or one can Plum tomatoes in sauce.
1 can Italian white beans, (canellini)
3 medium zucchini cut in 1 inch chunks.

Bring 6 qts water to boil and add ziti. Return to boil and cook al dente. *And I do mean al dente as it will cook more in the casserole.* Drain. Set aside.

Remove casings from sausage and brown it in skillet, stirring so you have loose meat. Drain off fat. Put meat in a large casserole, or soup pot.

Brown ground beef and garlic. Drain fat. Add meat to casserole or pot.

Brown sliced onions and stir into casserole.

Stir the tomatoes and the beans into the casserole, juices and all. Add zucchini and ziti.

Bake in 350° oven for 45 minutes. Serve with plenty of Romano grated cheese and crusty Italian bread to 'sop' up with. I add any leftovers too.

27

NOTE: If your family likes really "soupy" soup, add a cup of good red wine just before you put it in the oven. Also, the sausage seasonings are "hearty" so we do not add salt.

This freezes well by assembling to point of baking. I leave a casserole of this soup/stew in the freezer when I have to go out of town on autographing sessions. My husband and son can read my directions for heating on the label, which are same as above except double the baking time when preparing from the freezer. I even leave garlic bread wrapped in foil in the freezer with directions for heating. With ice cream or fruit for dessert, they make out pretty well!

However, and this is important, do remember my caution about freezing casseroles and sauces with *sausage meat* in them. 21 days is about the *limit*. I know I sound like a nudge but it is important.

ITALIAN POT ROAST
A variation of Baked Soup

An Italian friend makes this with stew meat cut in chunks about the size of a ping pong ball instead of ground meat and sausages. My Hungarian friend uses beef shank slices and swears the marrow from the shank bone makes it perfect. We are not fond of the extra fatty content from the shank marrow. But suit yourself.

However, we especially like to use a bottom round roast cooked this way, omitting the ziti. When using these cuts of meat, brown the chunks, shanks or pot roast first and allow to cook for about 40 minutes before adding rest of ingredients, then proceed. Serve the meat and vegetables over any starch, pasta, rice, or potatoes.

* * * * *

Let's just shove the soup pot to the back burner for a minute and talk about another stand-in for real meals. I am not a great fan of congealed salads, because those of my childhood were usually as sweet as sugar and I always wondered why they were called salads. But here is one that teases the palate as it nurtures hunger pangs. This makes a hearty chicken meal in an aspic.

A SALAD LUNCHEON ENTREE

2 cups chopped fresh peeled tomatoes, well drained, juice & seeds poked out and discarded.
3 cups cooked white meat chicken, 1/2 inch dice
1/2 cup finely sliced celery
1/2 cup sliced pitted black olives
1/2 cup thinly sliced baby sweet gherkin pickles
6 or 8 slender green onions, sliced fine
1 8 oz package lime Jell-o
1/4 cup V-8 or plain tomato juice

In large mixing bowl, toss first six ingredients together.

In medium bowl, prepare Jell-o, using 1/2 the amount of liquid called for, plus the 1/4 cup V-8 or plain tomato juice. V-8 really makes a statement!

Mix tomatoes and solid ingredients and the Jell-o together and pour into oiled or sprayed mold. I use a 13 by 9 inch glass casserole. Chill overnight or at least 6 hours.

29

Unmold on a bed of shredded hearts of iceberg head lettuce. **Easy Method:** Lightly oil or spray a cookie sheet. Place the cookie sheet over the mold. Hold tightly, turn over, food will now be on the cookie sheet. Now carefully slide molded salad onto platter of shredded lettuce. Frost the mold with the following dressing:
Whip together -
> *1 cup real mayonnaise (Lite works too).*
> *3 oz cream cheese (No fat if you like).*
> *2 Tbsp fresh lemon or lime juice.*
> *1 Tbsp grated lemon or lime peel.*
> *1/4 tsp powdered ginger.*

Options: Top frosted mold with 1 cup of salted sunflower seeds, cashew nuts, or smoked almonds.

When serving, be sure to serve up some of the crispy shredded lettuce too.

* * * * *

By now, especially if you have read my cook book, *How to Board Up Your Kitchen and Cook From a Hammock,* you know I love short cuts that make life as simple as possible. In the kitchen that often means "casserole." Meals in a dish were probably invented by some Neanderthal woman who did not have much in the way of serving pieces stashed away in her cave and did not have a four burner camp fire.

The French have a special way with them - cassuletts - casseroles, and we North Americans have developed a special genius when it comes to putting it all together in one pot. The following two ideas sort of evolved in my kitchen and have become favorites, because I am basically a pretty laid back, lazy cook.

SEAFOOD BROCCOLI BAKE

1 lb box linguini
1 bunch fresh green broccoli.
1 can cream of broccoli soup.
1 soup can skim milk.
3 Tbsp fresh lemon juice.
6 to 8 medium to large fresh ripe tomatoes
(you will need 3 layers of slices to fit casserole).
1/2 lb surmi (artificial crab meat).
1/2 lb cooked shrimp, shelled.
1/2 lb sea scallops.
1 bunch scallions, finely chopped.
3 sticks of hearts of celery, finely chopped.
1 1/2 cups shredded Monterey Jack cheese.
1 7 to 8 oz jar dry roasted pumpkin seeds
(pepitas). Sunflower seeds can be substituted.
Fresh parsley to decorate.

Boil the linguini according to package directions, al dente; drain, rinse in cold water and set aside.

Clean the broccoli, separate into small florets. Save the stems to slice into a salad. Blanch florets 3 minutes in boiling water and rinse in pan of ice water to stop cooking and preserve color.

Mix soup, milk, and lemon juice together using whisk to smooth.

Treat a casserole with non-stick spray.

Make a layer of overlapping sliced tomatoes in bottom of casserole.

Put a layer of half the linguini on top of tomatoes. Top with half of each of the three seafood items, half the chopped scallions and half the finely chopped celery, half

31

the broccoli and half the shredded cheese. Pour on half of the mixed soup, milk and lemon juice. Repeat the layers. Bake in 350° oven 40 minutes.

Cover with more overlapping thick slices of fresh tomatoes. Sprinkle the pepitas over tomatoes and bake another 15 minutes. Decorate generously with finely chopped parsley.

NOTE: I have also used 4 cups leftover chicken, or 4 flat cans of chicken in broth, or leftover baked ham instead of seafood. It is great either way.

You can say "TOMATAH - POTATAH," I don't care, but this is a vegetable and a starch all in one dish and sooooo easy! Should serve 6 or 7, with a meat and salad. Vegetarians can skip the meat, or add some crumbled tofu to each layer.

TOMATO POTATO

Preheat Oven to 400°.

Treat 10 inch ovenproof quiche dish with non-stick spray.

Create a seasoning oil by combining the following in a small mixing bowl.

1/4 cup canola oil mixed with 2 Tbsp Butter Buds.

1/4 cup chopped fresh basil.

1/4 cup chopped fresh Italian parsley.

3 rounded Tbsp grated Parmesan/Romano mixture.

1/2 tsp celery salt.

1/2 tsp white pepper.

Scrub 4 or 5 medium large baking potatoes.
4 or 5 large, ripe tomatoes. Wash, core and poke out seeds and excessive juice; slice 1/4 inch thick.
Slice scrubbed potatoes about 1/4 inch thick, and make overlapping layer to cover bottom of quiche dish.

Brush layer of seasoning/oil mix over potatoes.

Add a layer of tomato slices and more seasoning oil. We do not add salt, but if you do, do so sparingly.

Repeat layers of potatoes, seaoning oil and tomatoes.

Bake 40 to 45 minutes, until potatoes are tender when pierced with a fork. Add layer of shredded low fat mozzarella cheese.

Return to oven for just a few (4 or 5) minutes, till cheese melts but does *not* get crusty and dry. Serve at once.

* * * * *

Well, so much for good old American innovation in the kitchen...let's go back to those European band trips.

Ratatouille is found in some version or another all over France, Spain, Italy and Greece. So take your pick. Pronounce it by saying, "rat-ah-too-ay".

RATATOUILLE

1 medium, very shiny, firm fresh eggplant.
1 large zucchini, or 2 smaller ones, 1 inch dice.
Do not peel. (About 2 cups.)
1 large green pepper, cut in 1 inch squares.
1 large onion or 2 smaller, inch cube, petals separated.

33

2 Tbsp virgin olive oil.
Garlic, see note below.
1 pint Basic Tomato Sauce, thawed or 1 can
stewed tomatoes.
Juice of half a lemon.
Grated rind of half a lemon.
Pinch each of rosemary, oregano, thyme,

Peel and chop eggplant into 1 inch cubes. Place in deep glass bowl, sprinkle three tablespoons salt over it and toss it thoroughly. Let stand about 10 minutes then fill bowl with ice water and press and squeeze eggplant in water and drain water off. Rinse and press between paper towels. This treatment eliminates any bitter taste.

Stir fry zucchini, onion, peppers and eggplant in 2 tablespoons cooking oil. If you stir all the while you are sauteing, you won't need to add more oil, if tempted, add bouillon or just water.

NOTE: Now garlic is up to you - we add about three - yes 3 - *tablespoons* minced fresh garlic to the saute pan. *Don't scorch the garlic!*

The vegetables should be not quite tender when you add the thawed tomato sauce, grated lemon rind, juice, and herbs. You can tell; we love our seasonings, but we are stingy with salt. Simmer only 'till hot. Serve.

Also delicious chilled as an appetizer with sesame crackers on side. Served this way, it is sometimes called "caponata", especially when you toss in some good briny Greek olives and a dash of red wine vinegar.

MAKE IT GREEK

You can please Greek Gods with this one. To the above ratatouille, just before serving, stir in *one cup feta cheese chunks* and *half a can of sliced, drained, pitted black olives.* This is wonderful served over pasta; also very good over brown rice or barley. Makes a delicious nutritious, meatless meal. But if your family thinks they haven't been fed if they don't have meat, start with *1 lb ground lamb or beef, sauteed and drained well,* then proceed per the above ratatouille recipe.

NOTE: We often add 2 or 3 extra peeled, one inch dice, tomatoes to extend the amount. Sprigs of fresh basil leaves add visual flavor. Without the meat, this too can be served chilled as an appetizer with crackers.

* * * * *

No self-respecting tomato book would dare omit recipes from south of the Rio Grande.

Back in the late 1930's my Aunt Clara Sisler had a chili and beer parlor called THE CALICO CAT in Fairmont, Missouri, on old Highway 24 between Kansas City and Independence. Truck drivers en route from Chicago to Denver made it a point to stop for her chili, no matter the time of day. They loved the way she served her special recipe.

Aunt Clara has gone to Chili Heaven, but the CALICO CAT is still there, even though the highway has been replaced by I-75 and the old neighborhood, Fairmont, ain't the same anymore. But the neighbors out there are still Heart of America folks, who love people and love food!

35

She would cook the meat separately, with a whole handful of chili powder and when you ordered Chili, you got a small side dish of red beans, alongside a bowl of her meat sauce.

When we have chili, whether for two or a crowd, it's a party. Set someone to making Margaritas, (virgin for the kids) and be sure you have cans of kidney beans, black beans and packages of yellow rice in the pantry so you can vary the menu. While we're in the pantry, in all fairness, I have to admit that a can of commercial stewed tomatoes will work almost as well as our frozen Basic Sauce. Me, I'm partial to my own stuff. But suit yourself.

BASIC CHILI

1 lb small diced or ground meat. Your choice beef, pork, chicken, whatever, or a combination.
3 pints Basic Sauce, thawed
1 heaping Tbsp chili powder. Or more!
*1 heaping Tbsp cominos seeds, (whole, **not** ground).*
1/8 tsp ground cloves.
*1 tsp cocoa (**not** sweetened chocolate. Trust me!).*
1 cup finely diced onion
1/2 cup each finely diced green and red peppers
*1 cup finely diced fresh tomatoes, **no seeds**.*

Brown the meat, skim and discard all fat. Add all of the rest of ingredients. Simmer until you can't stand it any longer without tasting. *Now,* adjust salt and pepper, add more chili powder if you like.

The longer it cooks, the better it gets. It must not scorch, so frequent stirring is essential.

Time tip: When I don't want to play pot watcher, I stick it in a 200° oven as soon as it begins to simmer. Then I only have to stir it once in a while. If I am getting ready for a dinner party I start in the morning to simmer, and then pop into the oven until dinner.

NOTICE: Do not add beans to this basic meat chili sauce.

NOW COMES THE FUN!

You decide - Serve it with red beans; black beans; yellow rice; very thin pasta; over corn bread; or all of the above, presenting each in separate bowls so guests can create their own platter. Have lots of diced raw onions and a bottle of green, hot pepper vinegar handy for Texas types. A bowl of grated cheddar cheese and a bowl of salted pumpkin seeds (pepittas) or salted sunflower seeds are also wonderful toppings. Even people who claim to hate chili, will love this. Maybe it is because they get to be real creative as they prepare their own bowl. What could be easier on Super Bowl Sunday? Let your guests bring desserts and their own libation and use disposable bowls and plastic spoons. Aunt Clara used to serve a big basket of hard-boild eggs to cool fiery tongues.

Leftover meat chili sauce makes a quick supper when used to top hotdogs, a cheesy omelet, any size or shape of plain boiled pasta, leftover macaroni and cheese, thick slices of toasted Italian bread, or a slab of hot corn bread. Half a cup mixed with an eight ounce jar of any soft cheese spread, heated, makes a great party dip.

Before we go back to Europe, and while our tongues are atuned to spice, let's visit Louisiana.

We have friends from Mississippi who spent half their childhood visiting Grandma in Louisiana. They make the best Gumbo I ever ate. However they work all day at it. But since I introduced them to my laid back approach to cooking they use my recipe, basically. The husband is the real cook in the family, so like all men in the kitchen, he adds to, subtracts from and sometimes I am not sure if the resulting pot is Gumbo or Jambalaya. Neither is he! Even regional cookbooks seem to travel from one pot to another when it comes to Cajun cookin'. Creole cooks shrug Jambalaya off as being Cajun. Cajun gourmands like their menus to reflect a bit heavier, more country flavor. But no matter what you call it, Louisiana food is so good you'll overeat every time.

GUMBO

Considered Cajun, as are red beans and rice. But don't argue with anyone about it. The discussion will get as heated as religion, politics and sex.

1 pint Basic Tomato Sauce.
1 can okra.
2 cups chopped fresh tomatoes.
OR *1 can tomatoes.*
OR *an extra pint of your Basic Tomato Sauce.*
1/2 cup diced green pepper.
1/2 cup diced onion.
1/2 cup diced celery.
6 cloves crushed garlic.
1 can chicken broth.

Simmer all the above together for ten minutes.

THEN ADD:

1/2 lb cleaned shrimp (use small shrimp, or cut large ones in half).
1/2 lb small bay scallops or diced larger ones.
1 cup cubed leftover ham or sausages. (optional)
Even leftover chicken or turkey adds to the flavor, just like your Aunt from Iowa's vegetable soup, almost anything goes.)

Simmer ten minutes more and now add 1 tablespoon file' and 1/2 cup minced parsley. Stir together for about 5 minutes over the heat. **Do not cook any longer or the shrimp will get tough.** Serve in soup bowls and pass the hot crusty French bread for dunkin'. Good over rice too, but we prefer dunkin' the bread.

SEASONING HINTS: My Mississippi friends are apt to add any or all of the following during the last twenty minutes of cooking:

1/8 tsp each of mace, ground cloves, allspice.

1 Tbsp fennel seeds.

Also, we all have Louisiana Hot Sauce, green peppers in vinegar and the salt shaker on the table for individual preferences.

You can also add oysters, crab meat, or bite sized pieces of any fish you have in the freezer or fresh from the family Isaac Walton's catch.

When you read the following recipe, you wonder what the heck all the fuss is about between Gumbo and Jambalaya; and, incidentally, I've been served both where a meat broth was used instead of tomatoes. Will somebody from Louisiana kindly explain!

JAMBALAYA
This should serve 6

2 Tbsp Canola or good olive oil. Those original Louisiana cooks used bacon or fat back drippings. Forget it! It'll get you a stern lecture from your doctor.
2 medium white onions, chopped.
2 medium green peppers, chopped.
1 large rib of celery, sliced very fine.
3 large buds of garlic, minced.
1 cup diced fresh tomato pulp, discard seeds and juice. Better even than our Basic Tomato Sauce!
1 small can of tomato paste.
2 dozen medium shrimp.
2 dozen oysters, small.
Sometimes we use scallops. Sometimes both!
2 cups cooked rice.

Add the oil to a large hot skillet or Dutch oven, with a lid. Add everything **except the seafood and rice.** Cook gently on low heat, one hour, stirring often enough to avoid sticking. Add shrimp, oysters and scallops, continuing to cook gently for ten minutes.

Add the cooked rice and mix together gently. Serve at once. My gang like to add finely diced fresh green onions at the table. When I want to show off, I rim a pretty platter with fresh watercress, make a mound of the rice in the middle, and pour the Jambalaya over the rice and let it puddle around on the platter.

40

CAJUN CHICKEN

1 pint Basic Tomato Sauce, thawed.
1 can okra and tomatoes.
1 Tbsp file'
1 chicken, disjointed.

Simmer first two ingredients for 15 minutes then add 1 tablespoon file' and stir.

File' is a Creole seasoner found with spices; add last - if overcooked it gets stringy.

Brown chicken and arrange in a casserole. Now - pour your cajun sauce over browned chicken pieces (skinned if your doctor says so). Bake at 350° about 40 minutes or until chicken is done. Serve with cooked white rice or roastin' ears and a big green salad. Cross your fingers behind your back and claim to be a descendent of Evangeline. When they ask for the recipe, be gracious, give it to 'em.

* * * * *

When it comes to rice, we've never met a rice dish we don't like. Spanish Rice, was a great budget booster in my grandmother's kitchen when the tomatoes were ripening faster than she could pick them, but she had to go to school with me to get the recipe! So alright already, we're back in mid-America! But since Spanish is one of the Romance Languages, and this dish is sort of an ethnic food, why not?

Besides, Spanish Rice was probably the first food that made me aware that all the countries of the world cooked and seasoned their foods differently than Grandma. When I was in elementary school back in Bloomville, Ohio, the

Home Economics classes cooked and served the school lunches. Since most of the girls would eventually marry farmers and raise a family, even if they detoured to college first, they were taught to really cook full meals and plan around crops and budgets.

One of the favorite menus at school was Spanish Rice, cole slaw with chopped apples, or Waldorf salad, with oatmeal cookies for dessert. I finally prevailed upon Grandma to get the recipe from the Home Ec. teacher. Of course, Gram would improvise depending on what she had on hand or could afford to buy. I guess that's how I learned to "create" in the kitchen.

SPANISH RICE

3 strips of bacon.
12 large ripe tomatoes, peeled and chopped.
1 large onion, or green onions chopped (1 cup).
1 large green pepper, chopped.
2 Tbsp ketchup mixed with 3/4 cup water.
3/4 cup raw rice.

Now, read carefully, because I'll be telling you a story as I give you directions.

Fry out the bacon, crumble, and set aside. Pour off most of the bacon drippings. If you are on a medically directed low fat diet, for heaven sakes, forego the bacon, use a teflon skillet and add a dash of smoke salt for the country flavor.

Saute onions and green pepper in the skillet. Add the tomatoes and the chili powder. Mix ketchup and water and add to skillet.

Stir in the rice and the crumbled bacon. Now, if it was just too hot to fire up the oven, or if Grampa had already taken the coal stove down for the summer, Gram would gently simmer this skillet full on her kerosene stove until the rice was cooked. But she did have to stand around, handy with a spoon to stir and keep it from sticking. I just pop it all in a casserole and bake in a 350° oven and sit me down in a rocker under the fan with a pitcher of iced tea or lemonade. Gram used the oven only in winter when it was standing at the ready, all fired up in the big old coal stove. Why do we keep calling them the good old days?

If you get a hankerin' for Spanish Rice when fresh tomatoes are higher than a kite at the market, use your Basic Tomato Sauce or canned tomatoes.

About rice and microwaves: Frankly, I can't see any reason to microwave raw rice. It doesn't save any appreciable amount of time and if you ever have a mess of rice boil up and over in your microwave oven you will never try it again. If you love gadgets, and serve a lot of rice, a rice steamer is one that earns its keep.

* * * * *

I just can't call a halt to our ethnic recipe hunt without one more good Italian entree. On many of the American Winds Concert Band trips through the heart of Europe, we were royally entertained by the city fathers after each concert. One memorable evening was atop a mountain just outside the city of Pordonone. We were served a very formal, nine course Italian banquet, each course served separately on fresh china and clean silver.

The most delightful dish was a pasta smothered in eggplant and gorgonzola cheese. It was served as an appetizer, before an antipasto, which was followed by a veal entree, which was in turn followed by a simple macaroni in marinara sauce and then an icy, crisp mixed salad. There was also a wonderful cart of Italian ices, canneloni, tiramisu, fresh fruits, and rich ripe cheeses to accompany robust coffee. I am sure we all used up our supplies of stomach aids that evening.

We all voted this dish our favorite and, together, most of the wives managed to figure out the ingredients. We now serve it as an entree at our get-togethers.

This recipe takes a bit of time and is a mite rich, but it is absolutely delicious. I know it may sound strange but give it a try. You will love it and the "rich" won't hurt you just once. So splurge! Run a mile tomorrow and have a bowl of oatmeal for breakfast to unclog your arteries. Hide the Parmesan shaker, don't let anybody spoil the wonderful taste of the heated Gorgonzola cheese. Superbo!

PASTA WITH EGGPLANT GORGONZOLA

1 fresh medium sized eggplant.
6 or 8 medium sized ripe tomatoes, peeled,
seeded and wedge-cut into eighths.
(See Show-off Tip below)
1 large green pepper, sliced into thin strips.
1 large red onion, sliced into strips (a
white one will do, but the color is nice).
3 to 5 large buds of garlic - you decide.
3 Tbsp good olive oil.

First you peel the eggplant and cut it into 1 inch chunks and press and rinse in saltwater.

While the eggplant is in the saltwater, gently saute the garlic in the oil, watching it very closely so as not to scorch even a teensy bit. Then discard the garlic.

Now press the eggplant dry and quickly stir fry it in the garlic flavored oil. Keep it moving or it will need lots more oil, which *we* don't.

Add the rest of the ingredients and continue to stir fry until all vegetables are al dente. Meanwhile, have somebody at your side boiling linguini until al dente, and draining it.

Pile the pasta into a huge deep platter and spread the eggplant mixture over top. Now, the *piece de resistance* - crumble 3/4 pound of good quality Gorgonzola Cheese over top. Carry to table and accept the applause. Then, not before then, toss it all together and say, *"Mangia, mangia!!"*

We like to serve it in wide country soup bowls and have piping hot crusty bread to soak up the juices. Though there is no meat in this wonderful meal, a hearty red wine settles that rich gorgonzola down to a manageable richness.

Show-off tip: If you have access to fresh, vine ripened solid plum tomatoes, use them, cut into quarters and not peeled. There is just something exotic about plum tomatoes, romantic even!

Seems to me, by now you ought to be just bustin' to invent a casserole of your own. Remember, use up leftovers.

Speaking of leftovers - a bit of family fun. A friend had brought us a fresh caught dolphin fish to cook. (NOT

FLIPPER!) There was one piece left on the platter when Tom, my oldest son, said, "Somebody please eat that fish or Mom will throw a box of raisins at it tomorrow and call it bread pudding." My husband responded, "Oh no, she can make a good soup out of it. With a touch of nutmeg."

Now this wonderful man can't boil water but he is absolutely certain that nutmeg makes the dish, from fish to figs. But I did try one of his ideas one day when there were 2 small baked sweet potatoes left over and he suggested soup, "...with a touch of nutmeg." Yam soup? Well, why not. Know what? It was really good. I like it chilled best, but not him. The idea of cold soup makes him gag. Comes from being raised in Canada, I guess, where hot soup was synonymous with The Maple Leaf, The Queen and Mum's knitted mitts. So look for Doug's sweet potato bisque in the index.

STARTERS

Like I said back yonder a few pages, most cookbooks start out with appetizers. But I get around to these after I figure out what's what at the market and in my freezer to plan for the main event.

Whether you call them snacks, canapes, appetizers, or that French word nobody can spell or pronounce we all seem to thrive on a little "gnosh" before dinner, especially if it is a celebration for a crowd.

A good old Victorian starter was so simple and easy it was served first at tea time, and even the kids loved it. It was simply rounds of fresh white bread, spread with butter, sometimes seasoned with chives or cracked pepper and topped with peeled, paper thin slices of tomatoes.

46

Then of course, it was a natural progression to using delicate cucumber slices, replacing butter with softened and seasoned cream cheese.

Then one day, some inventive lady cut the crusts from a slice of bread and buttered it and wrapped it around an asparagus stalk. Boy, we were really getting fancy! And us frugal types cubed the crusts and made croutons instead of buying them.

Call it cocktails, or tea if your mater was from Britain, or tapas, as the Spanish do, here are a few little "appeteasers" that won't necessarily break the bank, send your cholesterol into space or run you ragged after a hard day. Naturally, the main ingredient is tomatoes (except for my tortellini dippers).

The following elegant appetizer can turn into a casual meal, or salad for guests to assemble to their taste. A bit lengthy, but worth it. This goody originated with my right hand, Ginger, who helped me get many a party off the ground. What's more, your guests can help in' the preparation. It's a fun kitchen caper, but beware of taste snatchers! *Be sure to read the recipe all the way through before starting.*

SEA AND TREE APPETIZERS
(Shrimp, Avocado & Booze)

First you make a really spiffy dip:
2 or 3 large tomatoes, chop, poke out juice and seeds so you have about 2 cups of tomatoes.
1 Tbsp brown sugar
1 tsp dijon mustard
1/2 tsp celery salt
*1 cup real mayonnaise, **not salad dressing.***

47

1/2 cup good quality Cognac liqueur (Ginger recommends Hennesy).
1 tsp lime or lemon juice.
While you're at it, squeeze about 1/2 cup lime or lemon juice into a clean spray bottle. (Read ahead!)

In a small saute pan cook tomatoes, sugar, mustard and celery salt on medium heat until they reduce to about half volume and thicken. Chill quickly by putting in a measuring cup in the freezer of your fridge. When this is cold, thoroughly mix with mayonnaise, Cognac and 1 tsp lime or lemon juice. Chill in a pretty bowl to place in the center of a large serving platter.

NOW, YOU NEED:
3 lbs large shrimp, cooked, veined, shelled but leave tails on.
1 or 2 very large Florida Avocados or 3 or 4 small California Alligator pears; ripe but not soft. (Read ahead about keeping them from darken ing.)

You may purchase already cooked cocktail shrimp from seafood market, or prepare them from scratch. If you cook them from scratch, bring 4 quarts of water to boil in large pot, adding 2 tablespoons of pickling spice to water. When at a boil, drop shrimps, shells and all in, bring back to boil, cover pot and set off fire immediately. Let set ten minutes. Shrimp will turn a lovely pink and will not be over cooked which is important.

Drain, cool, peel and devein, leaving tails on for "handle". Chill.

About the avocados, it is important that they be ready to eat, firm, not hard, and definitely not soft. Peel them, and using a small melon baller, make avocado balls. If time is of the essence just cut into cubes.

NOW, THIS IS IMPORTANT, and why I told you to read ahead. The avocado will turn dark unless you dip it in lemon juice, which is a tedious task. I keep a small clean plastic spray bottle in my kitchen and use it to spray bottled lemon juice over bananas, avocados, apple and peach slices, etc. All cut fruit stays pretty if given this "facial."

Arrange avocado around one half of a platter, place the Cognac dip in the center and arrange the shrimp around the other side of platter. Garnish as desired, bits of curly endive, Bibb lettuce leaves, olives, whatever turns you on. Have cocktail picks handy.

Those of us who have prolific avocado trees in our yards are forever trying to find a new way to serve these huge Florida beauties. Since avocados are also rather high in fat content, for years we have devised ways to dress them up without adding calories or cholesterol. However, now, we are being advised that the avocado oils are the good kind which increase our HDL levels. They make a perfect taste combination with our fresh Gulf shrimp.

The dip is the result of trying for a new taste instead of the same old tartar sauce or ketchup/ horseradish combo for shrimp.

An alternative quick dip for shrimp and avocado is made by blenderizing a good commercial chutney with one tablespoon of real mayonnaise and three tablespoons of good Cognac, chilled before serving.

49

Just so you wont think we never travelled anywhere except the Deep South and the Heart of Europe, let me take you on a little Island Hop.

We lived on Andros Island, the largest Bahamian Island, for about three years and fell in love with the foods, the people and the balmy laid back way of life. We had a babysitter who could make the best Coconut Bimini Bread I ever put in my mouth. What's more, she used an old beat up Sears and Roebuck portable oven which she set over a burner on a coal oil stove, much as my grandmother used back in the 1920's in Bloomville, Ohio. Just proves good cooks will always find a way to get the vittles on the board!!

One of the things our dining room at the old Lighthouse Club at Fresh Creek on Andros used to serve daily was conch salad. I swear I could live on this delicacy, and had it for lunch almost every day. Rowena, our babysitter, taught me to make it before we moved back stateside.

Mock Conch Salad
Serves 6 as appetizer

NOTE: This is an easily available take-off on the real thing, Bahamian Conch Salad, which is made with very finely diced raw conch. You can sometimes buy it in chunks at your seafood market. But it is a real devil to chop and usually very costly. Sometimes your fish market man will chop it for you. The Surmi, or Sea Legs (usually made of Pollock and available in most seafood markets), which is used here is as tasty, though not as chewy.

1 pound of Surmi, or use real conch.
1/2 cup fresh lime juice.
1 cup very finely diced raw onions.
1 cup very finely diced celery.
1 cup very finely diced green pepper.
2 cups of Salsa
(homemade or commercial, hot or medium, your choice).
Enough lettuce leaves to line six compotes.

Chop the Surmi about 1/4 inch dice. Toss with the lime juice, then add the rest of the ingredients - *except lettuce leaves.* Chill at least 6 hours, or overnight.

NOTE: If using real conch, allow the chopped meat to marinate all by itself in about a cup of fresh lime juice for at least 6 hours. This cooks the critter. Drain conch and proceed with the recipe.

Serve in wide mouth champagne glasses or stemmed compotes lined with lettuce leaves with a basket of torn chunks of Italian or French bread to "sop" up the juices. Have plenty of ice cold "libation" ready, (Mexican beer or Margaritas are perfect.) I usually serve this around my coffee table in the living room before dinner. It is a perfect lead-in to a cheese fondue supper also served at the coffee table. After the fiery salad and sufficient margaritas, the trip to the dining room seems formidable! This can also be a hearty topping for hot pasta.

Now here's a hot, off-the-shelf dip for veggies, crackers, shrimp, or chicken wings.

HOT STUFF FOR DIPPING

1/2 lb margarine container of Basic Tomato Sauce, thawed OR
1 cup canned stewed tomatoes.
1 7 oz can jalepeno peppers.

Blenderize about 1 minute. Put in top of double boiler over simmering water.

Add 1 pound grated Monterey Jack cheese or a 1 pound container of pimento cheese spread.

Heat and stir 'till cheese is melted and all is well blended. Serve in fondue pot with large corn chips, tortilla chips or chunks of bread for dipping. Have ample libation to cool fiery tongues and throats.

While tortellinis are pasta, usually served hot with sauces, we have found a zillion ways to serve them. Here's an easy goody.

TORTELLINI DIPPERS

1 lb box dried tortelllini, or use those wonderful, larger ones that come sealed in a vacuum pack and are just like fresh homemade. I really like them best but they do cost more. So splurge a little.
4 quarts boiling water.

Pour tortellini into pot and boil briskly until just barely tender (about 9 minutes for dried, 5 minutes for fresh).

Drain well, toss with just a smidgen of olive oil so they will not stick together. Chill. Serve on a platter with above Hot Stuff dipping sauce, or a dip made with 1/2 cup real mayonnaise and 1/4 cup dijon mustard. Supply fancy toothpicks or canape forks.

* * * * *

When you don't want an alcoholic beverage but do want something lippy smacky, no need to hit yourself in the face and say "Wow, I could have had a V-8" - just read on. I use this as an all day snack when I want to back off from a weekend of high caloric splurging that has shrunk my belts.

WONDERFUL FRESH TOMATO JUICE COCKTAIL

4 lbs ripe tomatoes, peeled, seeded and coarsely chopped.
2 cups very thinly sliced onion.
3 ribs very thinly sliced celery.
Dash cayenne or hot sauce, to taste.
Salt to taste.
2 Tbsp fresh strained lemon juice.

Bring all ingredients to boil. Put through food mill and then strain through fine sieve. Chill.

A VERY QUICK CANAPE

*2 large ripe tomatoes, peeled, cored, drained and
diced 1/4 inch.
1/4 cup each minced onion and green pepper.
1 8 oz block of lite cream cheese.
2 Tbsp real mayonnaise.
1/2 a packet Hidden Valley® Ranch seasoning.*

Thoroughly mix all ingredients. Chill and serve
with crackers or party rye bread rounds.

* * * * *

Now, come on, look at all those tomatoes; dream
up some canapes on your own. Try little slices of buttered
party rye topped with sliced tomato, paper thin slices of
onion and round slices of cheese - mozzarella or
Monterey Jack, whatever, and run it under the broiler in
the toaster oven at the last minute.

OR heat that fresh homemade Tomato Juice
Cocktail above in a pretty tea pot, pass it with a crystal
decanter of bourbon and your collection of fine china tea
cups. My Aunt used to call this her "Scarlet Sister Mary",
meaning no sacrilege, I assure you. It warms the cockles
of your heart on a snowy night, and does double duty as
cocktail and soup, and fools your tee totalin'
grandmother.

* * * * *

Speaking of soup, God must have made tomatoes with soup in mind. Because almost any good soup, that is not cream based, needs tomatoes. So here we go for some that can stand in as a luncheon, an elegant beginning, or a sturdy peasant supper entree. Soup can even be a nutritional, fast breakfast.

Whole nations of people around the world have stretched their rations with a good soup. When unexpected guests show up, soup to the rescue. It can be made fresh from scratch, contain leftovers from fridge, or a mixed bag from pantry shelves. But anything you simmer up in the stock pot is always better for the addition of a basic tomato sauce base, unless, of course it is a thick creamy chowder. And that's another book.

Try these on your tomato patch. This soup is really so good you will memorize the recipe and use it often.

TOMATO SOUP TO MAKE CAMPBELL'S JEALOUS

Puree 1 pint Basic Tomato Sauce, thawed, in blender with
1 Tbsp chicken bouillon crystals and
1/2 cup water. Heat thoroughly.
Stir in one cup no fat sour cream substitute or plain yogurt just before serving.
Top each serving with sprig of watercress.
Makes four seven ounce servings.

Add salad and crusty bread for a hearty lunch or light supper. Now that is about as easy as it gets. But read on, we get fancy without sweating over the stove all day.

PUREE OF MONGOL

Blenderize 1 pint container Basic Tomato Sauce, thawed, place in sauce pan and stir in 1 can split pea soup and 1 can half water, half milk. Heat and serve.
A little pile of sour cream on top is nice and a dash of nutmeg makes it special (I didn't need my husband to tell me that!). And a dozen thawed green peas add crunch and color. Don't cook them, just run boiling water over them...yummy.

Back home in Bloomville, Ohio, everybody's Grandma made vegetable soup and mine was no exception. Following is my version, good as Grandma's. After all, I'm a great grandma so guess I qualify as a vegetable soup maker.

OLD FASHIONED VEGETABLE SOUP

First, you take a bag of bones, from the butcher or from your freezer. Personally, I never cook a bone I can remove from a roast, steak or chop before cooking. I cut 'em out and freeze 'em. I even freeze them after I have cooked them, unless, of course, we have gnawed them!
Boil several bones and any left over meat scraps in about 4 cups of water, with 3 bouillon cubes, (chicken, beef, whatever) Skim as necessary. When meat falls off the bones, shred into the broth, give the dog the bones to clean his teeth.
NOW to this broth and meat, add:

56

2 pints Basic Tomato Sauce OR
2 cans stewed tomatoes.
1 cup diced onion.
1 cup diced celery.
1 10 oz box of mixed frozen vegetables, unless you
have lots of little margarine tubs of frozen leftover
veggies in the freezer. Use up whatever is on
hand, for heaven sakes.
HOWEVER if using leftover veggies, add them
after rest of ingredients are fully cooked.
1 zucchini diced.
1 very large or 2 smaller potatoes, diced.
*1 wedge of cabbage, (a cup coarsely **chopped**) not*
shredded. I hate trying to capture long shreds on a
spoon, and the kids will surely slurp and drip
cabbage all over the place. So chop!

Bring to a boil and reduce to simmer at once.
Cook until all vegetables are just tender.

TIP: *You can also omit the potato and add any of*
the following when it starts to boil: 1/2 cup raw rice, or
raw barley, or broken bits of pasta. Continue to cook on
medium heat until the grain or pasta you used is tender,
about 45 minutes for grains, and only about 10 minutes
for pasta or noodles.

OR use up leftover cooked grains or pasta too,
except don't boil, just heat. Adjust pepper and salt to taste.
Easy on the salt.

* * * * *

Permit me to digress a moment here and tell you about green peas in soup. You can add some to the above soup if you wish, but first - read on. When I was a kid, I hated canned peas, especially in soup where they got all mushy and ugly and usually cooked right out of their skins. Well I still feel that way about *canned* peas, but I love peas!

So when I want the taste of peas, or a recipe calls for them, I use a box of frozen ones. *And I do not cook them.* I dump them in a strainer, run hot water from the tap over them until they thaw. Then I add them to the recipe at the very last minute and allow them to get hot, but I don't cook them.

When adding them to a cold dish, I add them right from the colander as soon as they thaw. If I am making a salad calling for peas, the day before serving, I just rinse them from the frozen state and add to the salad. They will thaw overnight and are garden delicious! Okay, on with the soup section.

MINESTRONE

Start by assembling all of vegetable soup recipe above - *except* grains, peas and pasta. Then add:
1 can garbanzo beans, drained.
1 can red kidney beans.
1 can any white beans. (Try Canellini)
8 to 10 buds peeled fresh garlic, pressed
right into the pot. (No, this will not be too
much.)
Cook about an hour or more on slow simmer. *Then* add grains and/or pasta. Cook until they are done.

Serve in wide soup plates over generous links of well grilled Italian sausage and have lots of crusty hot bread to dunk. Serve a nice Chianti. Place a platter of fresh fruit and cheese (try soft Gorgonzola) as a centerpiece and then eat it for dessert. Invite the neighbors. Give every one a tiny sharp fruit knife and a chilled dessert plate.

Now I know I said this before but I don't want you to forget it: *About Italian sausages,* mild, hot or sweet, - if you boil them in beer before you grill or broil them, the fat will go into the beer instead of into your soup. This is also a great way to handle sausages of any kind (Polish, German, Smoked) before grilling. It reduces the grilling time, they do not dry out, and the fat boils away.

If you like that wonderful, rib stickin', before dinner salad the Italians call Antipasto, you will love this version as a before dinner soup. As a matter of fact it can stand alone as a lunch or light supper.

ANTIPASTO SOUP

1 pint Basic Tomato Sauce or canned stewed tomatoes.
1 Tbsp minced garlic.
1 can chicken broth.
1 cup sliced black olives.
1 cup small dice fresh cucumber, peeled.
1/2 cup raw pasta, (get fancy, buy shells or curly shapes.) Or use what is on hand (short pieces).
1/2 lb hard Italian salami, julliennd.
1/2 lb provolone or mozzarella cheese, cut into very thin jullienne strips.

59

Mix the Basic Tomato Sauce, the chicken broth and the garlic in a sauce pan and simmer until garlic is soft. Add cucumber, simmer 5 minutes. Add olives and pasta and boil until pasta al dente. Serve in big pottery bowls over thick slices of Italian bread which are perched on a layer of shredded iceberg lettuce. Yes, I said lettuce, believe me, crisp lettuce under hot stuff is good. Thai cooks do it! Sprinkle slivers of salami and cheese on top.

We Elliotts like to have a shaker full of hot red pepper flakes and a bottle of red wine vinegar handy to jazz things up. Chunks of Feta cheese give it sort of a Greek twist. Grated Parmesan/Romano keeps it strictly Italian.

* * * * *

Some kind of bean soup is native to every country in the world, I swear. From black eyed peas in the American South to beans and ham hocks in Germany, soup is a staple. This version borrows from several cuisines, and it doesn't matter what you add, it'll be good.

PASTA FAGIOLE

Use either canned canellini beans or buy them dry in a cellophane bag and hang around the kitchen all day cooking 'em. You can guess I use canned.

2 cans canellini beans.
2 pints Basic Tomato Sauce or canned stewed
tomatoes.
2 cans chicken broth.

Bring all to a brisk simmer. We add several garlic buds and a chopped onion when it starts to bubble.

Then add 1 cup short cut macaroni and cook until pasta is just past being al dente. Serve at once. You can also add shreds of any kind of leftover meat you have on hand, but it is good vegetarian style. One Italian family we know keeps a bottle of olive oil with several cuts of raw garlic in the bottom and shakes this rich flavored oil over their bowls at the table. Salty bread sticks go well with this soup, dunk 'em, crumble 'em up, or just munch 'em on the side.

* * * * *

On our trips through the Heart of Europe with the American Winds Concert Band, playing for our supper as we traveled, we touched into the roots of many of our favorite dishes, as well as several family trees. We usually stayed in small Mom and Pop pensions in the countryside near the major cities. Since there were usually 50 to 60 of us, we really got to know our hosts as only about 6 to 10 of us could be accommodated by each family. The plain folksy food was always fabulous and we could hang around the kitchen and learn a thing or two.

Often the family would entertain after dinner with folk music and dancing and members of our band would whip out their instruments and improvise, from Broadway show tunes to Dixieland. There were no political axes to grind in those friendly family circles as we stretched our minds and hearts across cultural traditions! Music, food and humor are the great communication bonds all humanity shares across all country map lines, the world over. And soups, especially

61

soups with tomatoes as a base, were *always* served, no matter what the dinner entree. The Earth could continue in it's orbit on the jillions of hearty soups that scent up kitchens from Albania to Zanzibar.

AUSTRIAN PANCAKE SOUP

Now, just a minute, I don't boil pancakes up with bones! Hang in with me.

For this you will have to make about 6 thin crepe like pancakes and sliver them into thin strips and have at the ready. I like to go on a crepe spree, make about a dozen or more up when I'm in the mood, stored in freezer between layers of wax paper where they are ready for a dessert or this delicious soup.

CREPES
Should make 8 to 10

1 cup milk.
3 eggs.
1/2 cup flour.
1/4 cup canola oil.

Put all ingredients in blender, mix on high, scrapping down sides, until well blended. Let sit at least half an hour. Spread by 3 tablespoons at a time over hot non-stick sprayed skillet. Turn almost at once. Stack between wax paper until all batter is used. Can be frozen or used at once. Can also be used to wrap around fruits and or creamed chicken or seafood.

BUT, FOR PANCAKE SOUP -

1 pint Basic Tomato Sauce, or 1 can tomatoes, OR
2 cups chopped fresh tomatoes and then proceed
with the following ingredients.
1 can each beef and chicken broth.
1 cup sauerkraut, re-chopped so not
stringy, and rinsed in a strainer with very
hot water to cut the tart.
Note: *you can use shredded fresh*
cabbage instead.
1 large carrot, grated.
1 cup each very finely diced onion & celery.

Bring all to a simmer until kraut (or cabbage) and carrot are tender.

Ladle into soup bowls and top with generous bundle of jullienned sliced crepe strips. You will need pumpernickel or caraway seed rye bread with this. This soup is served all over Germany and Austria with slight regional variations as a lite start to most meals. We enjoyed it so much we often asked for large bowls and plenty of bread and made a meal of it. A good cold German Beer is great alongside. Or a nice Rhine wine. We occasionally found these crepe strips in plain rich broth. Still delicious.

BARLEY SOUPS

We found soups with barley all over Europe and most had lots of tomato chunks and a heavy onion and celery flavor. To a pint of Basic Tomato Sauce, add a can or one pint of broth, a cup or so of leftover meat and a cup

63

of barley. Cook until barley is tender, about half an hour. You can add almost anything from the vegetable patch to this.

Some basic rules about creating soup

A half cup of barley and two cups of any broth will extend soups of any kind.

To a pint of Basic Tomato Sauce and 1 can any kind of broth add any one or several of the following: Broken odd pieces of pasta (that leftover lasagna noodle you didn't need), tortellinis, raviolis, gnocci, any leftover vegetables, cooked or from the crudities platter last night, shreds of leftover meat - got the idea? So go make soup.

* * * * *

Now then, remember way back in the beginning of the book (on page 5), we suggested freezing some Basic Tomato Sauce in those *half pound* margarine tubs? They are just right for salad dressing since you really don't want to make it up in large quantity.

SALAD DRESSING #1

1 half pound margarine container of Basic Tomato Sauce.
3 buds garlic, peeled.
1/4 cup tarragon vinegar (Plain vinegar will do, add a pinch of dried tarragon).
1/4 cup canola or virgin olive oil.

Blenderize, chill and use on any vegetable salad.

SALAD DRESSING #2

Blend into above recipe 1 cup yogurt & 1/2 cup
real mayonnaise, *not cooked type salad dressing.*

Chill well for at least an hour before using. This is delicious on steamed, chilled veggies, such as brussels sprouts, tiny carrots, asparagus, broccoli and raw or canned whole plum tomatoes. Steamed and dressed veggies make a wonderful salad when lettuce and tomatoes are not in season. You can also add canned artichoke hearts to your salad.

SALAD DRESSING #3
MADE WITH FRESH TOMATOES

6 large, really ripe tomatoes.
1 tsp Jane's® Crazy Mixed Up salt.
1 tsp lemon pepper.
1/2 cup salad oil.
1/4 cup tarragon vinegar.
1 Tbsp sugar, to taste, to cut tartness.
3 peeled garlic buds.

Peel, stem, core and poke seeds out of tomatoes and dice into blender container with rest of ingredients. Liquefy. Store in fridge. Dieters may omit the oil, but the amount used breaks down to very low fat and calorie count per serving.

Believe it or not, their are salad lovers who hate garlic. So the following fresh tomato salad dressing is right up their alley.

65

TOMATO SALAD DRESSING #4

Peel, seed and coarsely chop 2 good sized ripe tomatoes, place in blender with the following:

2 tsp dijon style mustard.
1/2 tsp salt.
1/4 tsp crumbled dried tarragon leaves
(or 2 tsp minced fresh).
2 Tbsp wine vinegar.
1/2 cup extra virgin olive oil.

Puree, all ingredients except olive oil. Then, with blender set to emulsify, slowly add olive oil until dressing is creamy. Add pepper to taste. A tablespoon of grated shallot or finely minced green onion is good added at time of serving.

* * * * *

Now, about those really ripe tomatoes from your U-Pic caper or your own garden.

TOMATO PRESERVES (A Great Gift!)

4 cups diced, peeled, and mostly seeded
tomatoes, drained of their juice.
1 can tomato paste (6 oz size).
1 can crushed pineapple in natural juice (16
oz size).
1 box powdered pectin.
1 lemon, thinly sliced and seeded.
5 cups sugar.
1/8 tsp each powdered cinnamon & cloves.

In a large pot, cook the tomatoes 'till the juice starts to diminish. Stir in the pineapple, pectin and lemon slices. Bring to full boil and stir rapidly about three minutes. Add sugar and boil, stirring about 5 minutes. Cool about 5 minutes, stirring to keep the fruit from floating. Pour into hot sterilized jars and seal with paraffin and lids. Makes about six 8 ounce jars full, with a little dish left over to sample right now. Store in fridge.

* * * * *

When barbecuing or broiling meat, chicken, or seafood, tomatoes seem to be the number one ingredient for sauces. Even supermarket deli ribs or chickens can be jazzed up by adding three tablespoons of your favorite bottled barbecue sauce to an eight ounce margarine container of your Basic Tomato Sauce. Heat together and brush on your deli selection at the table.

Speaking of jazzing things up, the following marinade works beautifully with flank steak, London Broil, Porterhouse, or even thick cut California or chuck roast.

JAZZY STEAK MARINADE

1 8 oz bottle Italian Vinaigrette Salad Dressing.
1/8 tsp Wright's® Liquid smoke.
3 Tbsp Lea & Perrin's® White Wine Worcestershire Sauce.
1/4 cup white wine.
2 buds garlic, peeled and split.
1/4 cup white wine.

2 cups peeled, chopped fresh tomatoes OR
2 cups well drained canned chopped tomatoes.
2 cups chopped sweet onions.

TIP: If your onions are strong enough to pull your eyes out, sprinkle with 1 tsp salt, 1 tsp sugar, and 3 Tbsp water. Mix and crush together, let stand about 3 minutes. Drain, use onions as directed.

Blenderize all ingredients until fully emulsified. This marinade will keep in refrigerator for ten days. Use as follows:

For a Flank Steak: Snip off fat. Score diagonally on both sides with a sharp knife. Place in flat glass casserole. Pour 1/4 cup of marinade over steak. Rub in thoroughly, turn steak over, continuing to rub in to saturate the meat. Add marinade sparingly as needed. You don't want to drown it, but it must be fully soaked. Store in fridge 6 hours or all day, turn and rub occasionally.

To grill: Mix 1/4 cup white or red wine with 1/4 cup marinade from the casserole, heat and keep hot while you grill the meat four inches from heat, six minutes on first side, four minutes on second side for rare. Adjust time by the minute for degree of preferred doneness. Remember that well-done flank steak might be tough. Slice steak thinly, slantwise across the grain, arrange on platter and pour heated wine marinade over.

Top Round London Broil & other steaks:
Poke full of slits and holes with cooking fork and thin boning knife. Marinate and rub as for flank steak. A three-inch thick London Broil will need to be five inches from heat, four minutes at a time on each side, turning about three times for rare, Medium rare, increasing turning times, will still be good, but too well done will be tough. To serve, slice same as flank.

California & chuck roasts: Make slits, poke holes, rub and marinate as directed for steaks. Pot roast slowly in a 300° oven until well-done and tender. Wonderful sliced thinly and served on big, fat onion buns.

With the steaks, have a crusty loaf of Italian bread to tear into chunks and dunk in the marinade. Mixed green salad and steaming ears of sweet corn make this a meal fit for special guests and easy to prepare. The zesty taste of the simmered tomatoes and onions in the wine compliment the buttery, country taste of sweet corn. Do not repeat the tomato taste in the salad; use assorted greens, cucumbers, green peppers and celery for crunch. This is called balancing taste sensations. This slight of hand with your palate satisfies hunger before you have stuffed yourself beyond comfort; a good rule for any meal. So is chewing each bite at a time, slowly!

* * * * *

When it comes to a basic barbecue sauce, the following recipe is equally good on ribs, lamb shanks, chicken, or seafood.

ALL-PURPOSE BARBECUE SAUCE

5 pounds of tomatoes, peeled & chopped.
1 pound onions, diced very fine.
1 16 oz. can crushed pineapple in natural juice.
2 heaping Tbsp brown sugar.
1 tsp powdered smoke salt. (Buy at spice counter.
You may also use Wright's® Liquid Smoke, but
use only 1 teaspoon, don't be tempted to squirt on
a dash more. It is powerful stuff.
1 tsp white pepper, (adjust to taste)
1/2 tsp celery salt.
1/2 cup cider vinegar.
1 tsp garlic powder. OR 2 or 3 crushed garlic
buds, (We prefer 5 or 6 buds).

Bring all to boil and then simmer and stir often for about 1 hour, uncovered; it will get nice and thick. For some reason, it never makes the same amount twice. Store in fridge in sterilized pint jars. Makes great hostess gifts.

USES - As a marinade: add 1/2 cup burgundy wine to 1 cup sauce. Marinate fish or fowl 1 hour in fridge. Marinate pork, beef, or lamb 6 hours; or overnight in fridge for head start on a party. Also, brush on while grilling.

As barbecue sauce, without marinating, brush on any meat while grilling, but do turn often so the sauce won't burn before your meat is done.

In meatloaf, four tablespoons of sauce mixed in meatloaf mixture makes it company fare.

As a condiment, pass a cup of heated sauce at the table when serving burgers, hotdogs, or ethnic sausages.

When I was a child growing up in Ohio, barbecuing was not usually done by my family, but was a joint effort for family reunions or village celebrations. However, my grandmother made her hamburger patties extra-special by serving with the following recipe. She would first sear the patties in a hot black iron skillet.

TOMATOES IN CREAM SAUCE

Peel and slice about 6 large tomatoes or 8 smaller ones. Arrange slices in Corning Ware® skillet or other stovetop cook-and-serve ware; or break out your electric skillet and cook on the back porch.

Drizzle about 3 tablespoons melted butter or margarine over tomatoes (don't count calories tonight). Now, gently saute about 5 minutes. Sprinkle 2 tablespoons minced watercress and 2 tablespoons parsley over. Keep sauteing. Sprinkle on a heaping tablespoon Wondra Flour® and 2 heaping tablespoons fresh basil. NOW pour a pint of half-and-half (cream) over it all. I did warn you about the calories, didn't I? Simmer 'till nice and thick. Serve it over hamburger patties, mashed potatoes, hot biscuits, noodles or linguini; or just in a sauce dish with a spoon. Diet tomorrow, or use skim milk & butter buds in recipe instead of cream and butter.

* * * * *

On our band trips, we learned that most European families serve salads after the entree, even though glitzy city restaurants follow the North American custom of keeping diners content with the salad while waiting for the main entree. Also, we were often served delicious salads that were not based on lettuce or other greens, but used a wide assortment of other garden vegetables.

I began cooking for a family in 1939, and salad greens were scarce and expensive in the winter, as were fresh tomatoes. But I recalled my grandmother's winter savories, which I now realize she must have learned from her Scottish, French, and Irish ancestors.

OFF-SEASON SALADS

1 can storebought whole tomatoes in juice. Drain and save the juice for the dressing.
1 can artichoke hearts, quartered (discard juice).
1 box frozen broccoli cuts, steamed al dente, chilled. Arrange all veggies in bowl.

DRESSING:
Add the saved tomato juice to:
1/4 cup oil.
1/4 cup wine vinegar.
Pinch each of tarragon, oregano, basil.
1/4 cup sliced green onions.
1 tsp sugar.

Whisk together 'till well blended. Pour over vegetables, decorate with parsley, and chill thoroughly. Makes a pretty, edible centerpiece for a Christmas buffet table, especially in Grandmaw's cut glass berry dish.

VARIATIONS
Starting with 1 can of tomatoes and the above dressing, any combination of the following vegetables make a wonderful off-season salad. Adding cubed cheese and leftover meats will give you a luncheon entree.

Canned baby corn.
Garbanzo beans.
Flat Italian green beans.
Kidney beans.
Canned hearts of palm
Canned sliced beets.

Frozen peas, thawed.
Sugar snap peas, thawed.
Pea pods, fresh or frozen.
Broccoli and/or cauliflower
raw or steamed & chilled.
Assorted olives & pickles.

* * * * *

Today, in our markets, we are confronted with multiple choices of relishes, and condiments. Well, Grandma made a jillion versions of these zesty enhancers and called them savories. I bet your ancestors did too, no matter where they came from. Whenever my Grandmaw made the following fruit relish Grandpaw would just up and kiss her right in front of us kids.

This relish is good alongside any roast or fowl. And if you mix a couple tablespoons into an 8 ounce block of cream cheese it becomes a dip or spread for toast points or crackers. Grandpa used to spread it on hot homemade bread if there was an open jar when the bread came out of the oven. Not me! I do have my limits. But I do remember that a neighbor who used to bring us wild game, always asked for a spare jar in trade. Seemed like a fair exchange to me. I can't really decide if it's a savory or a sweet. No matter, it's real good with rabbit, squirrel, or venison, and especially good with roast pork.

Before we go on, let me make a suggestion about storing homemade condiments and sauces. Back in the olden days, women stewed and stirred, sterilized, and processed jars in gallons of boiling water. But they had

73

cold cellars instead of refrigeration. Today, I do not process jars of food in boiling water; nor do I do bushels and bushels at a time. Furthermore, we Floridians do not have basements. I find a second fridge with a freezer top more useful than a large second freezer.

FRUIT RELISH WITH TOMATOES

5 lbs ripe tomatoes, peeled, cored, one inch dice.
6 Anjou or Bartlett pears, 1/2 inch dice.
6 peaches, peeled, 1/2 inch dice.
4 large onions, very fine dice.
1 large sweet red pepper, cut into short jullienne strips.
1 16 ounce can crushed pineapple in pure juice.
1 cup raisins.
3 cups white sugar.
2 cups cider vinegar.
1 Tbsp salt.
2 Tbsp whole pickling spices tied in a bag.

Put all ingredients in large pot and bring to boil. Keep on an active simmer and stir lots and lots until it gets really thick. It might take 2 hours. Ladle into sterilized jars and seal with melted paraffin and lids. I store mine in back of fridge.

A friend who loves chutney, mixes a level teaspoon of curry powder into a half pint jar of this relish when I give her one for Christmas and her family love it. As for me, when it comes to chutney, there are only two in the world I really like, Major Grey's and the mango chutney my son-in-law's grandmother used to make.

* * * * *

No tomato book worth its salt would be complete without delicious ways to serve green tomatoes. If you have no shiny green tomatoes on hand, go pick some, or order them from your market. Most roadside vegetable stands will get them for you.

GREEN TOMATO PIE
Better than apple!

Start with your favorite pastry for two crust pie; or save time by using ready made crusts from the freezer section at your supermarket.
6 or 8 medium sized green tomatoes. Stem and core, wash well, do not peel.
Slice them thinly into a bowl. Now add:
2 Tbsp cornstarch and toss well.
3/4 cup brown sugar.
2 Tbsp lemon juice.
1 tsp powdered cinnamon.
1/4 tsp powdered cloves.
1/4 tsp nutmeg.
1/2 cup raisins.

Toss all together to mix well. Pile in pie crust, cover with top crust, making slits for steam. Bake at 425° for 15 minutes. Then at 350° for 20 minutes. Brush top with slightly beaten egg white or cream or canned milk and sprinkle 2 Tbsp white sugar over top. Return to oven for ten minutes, or until a lovely brown. Watch it, as it will burn easily. This is delicious cold, but served hot with vanilla ice cream melting over it will get you standing ovations. This was Grandpa's favorite dessert.

* * * * *

Now, I swear, we Midwesterners were frying green tomatoes long before Fannie Flagg found the Whistle Stop Cafe. I think we borrowed from the Amish and then Southern cooks learned from us.

FRIED GREEN TOMATOES

6 good sized green tomatoes, washed, stemmed, cored, but not peeled, sliced about 1/4 inch thick.
Dredge in following mixture:
1/2 cup Wondra Flour.
1 Tbsp corn starch.
1 tsp salt.
1/4 tsp pepper.

Place in single layer on platter and chill for at least 1/2 hour, then fry gently in 2 Tbsp oil and 2 Tbsp margarine, mixed. Turn when crusty brown and fry on other side. Do watch them, they scorch quickly. You can cook up a batch by keeping them warm on a heat proof platter in a low oven. If oil seeps on to platter, blot it with a paper towel before serving.

These are usually served as a side dish, but do try the following Elliott family invention.

FRIED GREEN TOMATO PASTA: To make a delicious entree, use fried green tomatoes to top a platter of cooked angel hair pasta and cover generously with grated parmesan cheese (Drop two garlic cloves in the water in which you boil the pasta). This makes a wonderful vegetarian meal with a mixed green leafy salad.

Now don't turn up your nose at these next two variations of old favorites. Try them, and I bet you'll always want to make them with green tomatoes.

This first one is from our colonial ancestors and the second from an Italian writer friend.

GREEN TOMATO SUCCOTASH

5 medium green tomatoes.
1 ten oz box frozen baby lima beans.
1 ten oz box frozen whole kernel corn.
10 single soda crackers, roughly broken.
1 stick margarine.
1/2 cup milk.
1/2 tsp salt.
1/8 tsp white pepper.
6 oz grated cheddar cheese.

Stem and core tomatoes, dice one inch. Melt margarine and mix with milk, salt and pepper. Toss all the vegetables with broken crackers and arrange in casserole. Pour milk mixture over all, add the cheese and stir together. Bake 45 minutes at 350°. Top should be brown and liquid thick. This should serve 8 to 10 persons as a side dish.

Please, when you read the following title, do not squirm and go, "yukk!" Just try it. If you hate it, write me and I'll send you a new recipe from my kitchen.

GREEN TOMATO CLAM SAUCE

6 green tomatoes, stemmed, cored, dice 1 inch.
1 can whole clams (I use Geisha®, 10 oz size).
1 can white clam sauce. (I use Progresso®).
3 cloves garlic.
1 cup chopped fresh or frozen parsley.
1 lb linguini.

Smash and saute the garlic in 2 tablespoons oil in sauce pot or large skillet. Discard garlic. Add tomatoes and saute 'till tender and limp. Add parsley, clam sauce and clams. Simmer very, very slowly while you boil the linguini to al dente stage. **NOTE:** the clams will get tough as bubble gum if you boil them.

Drain linguini, arrange on large, deep platter. Pour the clam sauce over. Use a fork to pull and distribute the clams and tomatoes equally. Sprinkle with lots of grated parmesan or romano cheese. Be sure to have a hot, crusty loaf of Italian bread to tear into pieces and sop up the sauce. With a hearty salad, this should fill up 4 hungry people. Keep lots of these ingredients on the shelf for spur of the moment parties.

Green tomatoes have even found their way into the pickle jar and Kosher cuisine. More years ago than I want to remember, a group of us young Episcopalian mothers were being trained to be Sunday School

Teachers. We would gather on a Thursday morning for Holy Communion and a training class in Bible and Catechism. At noon, our teacher, a young guitar playing curate fresh from Seminary would lead us all to a nearby pub. Yes, beer garden. But don't get excited, the attraction was the barrels of wonderful pickles and garlicky Kosher green tomatoes they set out on every table, along with their menu of famous "mile high" sandwiches, the minute you sat down. The fact that most of our husbands would join us for lunch from their nearby offices enhanced the occasion. There, we would discuss LIFE and sing a song or two to Father Kee's guitar strumming. How clever of him! He hadn't a prayer in those days (mind you, this was the early fifties) of getting any young husband to teach Sunday School, but oh how he exposed them to the teachings of the church!

And I drool now remembering those Kosher green tomato pickles I would slice into my rare roast beef sandwich. I make them every year and keep them in my fridge and even my Jewish daughter-in-law seems to like them. I often serve samples of them at my book signings.

I believe this easy recipe first appeared as a reader's contribution to the food section of the Miami Herald, many years ago. It has since been passed from person to person with all kinds of personal variations. This version works well for my crowd.

Jars of these can be sold at your church or PTA bazaar and would be welcome gifts to bachelors or your hostess when you're invited out to dinner. After their 48 hour curing at room temperature, they must be stored in the fridge and served icy cold. Be sure to read the entire recipe through before you start, as it is not in traditional recipe form.

KOSHER DILLED GREEN TOMATOES

FIRST you take a 32 ounce wide mouth canning jar, sterilize it well, and line the bottom with celery cuttings, leaves and stalks. Now add 6 or 9 large garlic buds, peeled, and a snippet or three of fresh dill. Don't worry, if fresh dill is not available, use a level teaspoon of dried
NEXT, you take nice firm medium sized green tomatoes, wash very clean and cut into quarters. Really scrub the stem spot!
PACK them into the jars. NOW add three to four ounces of white vinegar.
NOW, THIS IS IMPORTANT - To each 8 ounce measuring cup of hot, very hot, tap water, add 1 teaspoon salt. Stir until salt is dissolved. Fill the jar to the very top with this mixture, making additional cups full as you need it. TIP: If your gang likes very spicy stuff, put in about a teaspoon of hot pepper flakes too. Cover the jar with regular canning lids. Keep lined up on the kitchen counter for two whole days, 48 HOURS! Then refrigerate for two days before eating.

They will keep crispy fresh for months, refrigerated. But I bet you, once opened, your family will devour them immediately. A great gift from your kitchen!

BONUS: You can use this same procedure to make **Kosher Dill Pickles,** just use nice uniform size small, whole, very fresh, crisp Kirby® cucumbers instead of green tomatoes. Do not quarter the pickles. Just scrub. Another super gift from your kitchen. I love these, and the dill tomatoes, sliced onto generously buttered pumpernickel or a good Jewish rye bread. What a sandwich!

My husband has always accused me of eating with my eyes. So - I love candlelight and china on linen. I'm a sucker for food presented artistically on my plate, and waiters who seem to glide to soft music.

However, I think I eat with my soul. I find joy and fun attacking a picnic basket and slapping together an impromptu sandwich in the kitchen. Seems to me as I think about it, most of my fondest memories have something to do with food, family, and friends.

Following are some very special treats that bring back wonderful memories from my life. Isn't it lovely how certain scents and aromas can transport us to another place and another time? Hot, bubbling ketchup, simmering tomatoes, the lingering zest of onions being chopped by the peck. What a wonderful kitchen aroma it was. And what a busy day it was for the women who seldom took time for lunch during those hectic harvest and canning days. Grandmaw used to make this for herself in the middle of those days. I used to think it was yukky, but with age came appreciation. I love it now.

GRANDMAW'S SNACK

Thickly slice a nice big ripe tomato.
Sprinkle slices with just a little bit of brown sugar.
Top each slice with a dollop of sour cream,
(Gram's was in a crock, leftover from breakfast.
Today, I use the commercial variety.)

Now, Grandmaw ate it just this way. I often broil them for a couple of minutes before adding the sour cream. I also make these slices on a jelly roll pan as a side dish with ham or chicken.

One wonderful summer in Rapallo, Italy, we awoke every morning to the mixed scent of the sea beyond our balcony and a wonderful aroma rising from a small bakery beneath our window. I soon found my way there and watched as dozens of local people emerged with a huge square of flat bread in their hands, which they would eat as they went on their way to work. I had to try it. Following is the best re-creation of that wonderful taste that I could come up with in my kitchen at that time. Mind you, this was in the early '80s and now there are many ethnic cookbooks with several versions, and even our local supermarket bakery makes it. Well, here is my version, easy as all get out and reminiscent of an Italian village.

FOCACCIA

You can use either a tube of biscuits, or a tube of French bread dough, to be found in the dairy case (or use a ready baked Bobboli® crust - see below) Flatten dough into a square pan, joining seams. Saute a large thinly sliced onion in 2 tablespoons olive oil until limp.
Pour onions & drippings over the dough.
Then top with grated mozzarella cheese, about 8 ounces. I know it sounds like Pizza, but it's not.

Bake at 375° for about 20 minutes, or until puffed and dough is done. Eat right now, piping hot, with salad, soup, or just alone with a glass of wine. I swear it will make you speak with an Italian accent! **NOTE:** If using Bobboli® crust, 8 to 10 minutes in oven will do it.

Sometimes our weary feet stopped us in our tracks and another tasty European memory was born. On the band tour of Scotland and England, a group of us were bone tired from standing in line in Westminster Abbey. We were also starving. My feet were killing me, and I knew we were in for a heavy dinner later in the day. We just popped into the first restaurant we came to. It turned out to be a little Italian sidewalk cafe, along the banks of the Thames, with no real menu. I asked for a salad and some cheese and bread. What I was served has since become a regular in most Italian restaurants in this country but in 1986, it was new to this country gal.

TOMATO AND MOZZARELLA SALAD

1 lb smoked mozzarella cheese, cubed.
4 or 5 large ripe tomatoes, seeded and peeled, cut into quarters.
1 cup of fresh whole basil leaves.
3 Tbsp wine vinegar, or fresh lemon juice.
1/3 cup good fresh virgin olive oil.
2 or 3 cut garlic buds.

In small bowl whisk salt and pepper to taste with vinegar, (or lemon juice), basil and oil until emulsified.

Put the cheese and tomatoes into a salad bowl which has been well-rubbed with the cut garlic buds.

Add the whisked salad dressing and toss. Serve room temperature or chilled, your preference. This will serve 4 or 5 as a side salad on bed of lettuce. As a meal with bread, two people will lick the bowl, or at least wipe it out with the bread!

Our waiter kept refilling my plate from a huge icy aluminum bowl and our basket of hot bread frcm ovens that looked like they came out of the ark. I swear I was falling in love!

* * * * *

Romance can thrive anywhere. You've heard of shipboard encounters? Well, when our Italian tour bus driver and one of our Midwestern female horn tooters fell in love in Switzerland, we all participated vicariously for twenty-one days. Then their tearful farewell embrace as we boarded our return flight home from Frankfurt brought us all back to reality. However, we all admit to heightened and strengthened relationships after many of our individual encounters on those tours.

In Venice, on the Mediterranean Riviera, and in the palace and chateau country of France, romance is inevitable.

One day in Venice, another couple joined us as we played hooky from band rehearsal and prowled the back streets. The sound of a slightly off-key tenor, the raucous scolding of an angry woman and the aroma of garlic, sausages and cheese pulled us into a very narrow alley, one canal behind St. Mark's Square. The tenor was sitting on the pavement, leaning against a wall. A battered stringed instrument and a jug of wine shared his spot of shade. "Mama" was leaning out of a window overlcoking the alley gestisculating angrily.

Three rustic iron tables and half a dozen rickety chairs crowded the doorstep. A really primitive poster showing a hand drawn bottle of wine and plate of food indicated we might be able to get lunch. There was absolutely no chance of our fractured Italian and Mama's brisk questions in her native tongue ever bringing about an understanding. So we all pointed to the sign and indicated we would be happy with whatever she managed to bring us.

Wow! Talk about a meeting of minds! First she brought a jug of wine and four mismatched glasses. Then just as we were pouring our third glass of wine and had about given up on ever seeing food, she appeared with a basket of hot crusty bread and a platter of assorted cheeses. I drool at the memory of that melting gorgonzola, almost like cream, a smoky hard cheese we took turns shaving, and others we had never had before, each one better than the last.

About the time we decided that this was lunch, and indeed, it should have been - she appeared with a platter big enough to hold a 20 pound turkey, piled with pasta that had been tossed with about a peck of quartered fresh tomatoes, sliced onion rings, a motley assortment of steamed, chilled bits and pieces of seafood - clams, prawns, white fish,- we figured whatever was left in the kitchen from last night, or maybe gleaned from the morning purchases against the arrival of the dinner crowd later on. She also gave us a bottle of olive oil with about a dozen buds of garlic floating about in it.

The pasta was hot, the vegetables cold and she gave us each a lemon and a small sharp knife and showed us how to puncture the lemon and squeeze it over the pasta, and to shake the garlic oil on top, then toss together, before serving ourselves on mismatched crockery bowls.

Following the above description, we try often to make it in our kitchen, and come close, but the memories of that afternoon with Papa's music(??), Mama's scolding, the table that kept threatening to collapse and that food. Mama Mia!!

Read these pages again, and just take off on your own, with friends gathered in your kitchen offering suggestions and be sure someone thinks they can sing, or at least have a tape of Julio Eglasias handy. We use the basket of bread and plate of cheese to accompany fresh fruit *after* the pasta, instead of before, but "Mama Mia" kept us happy in that Venetian alley with her version while she boiled pasta and cleaned vegetables. Do it your way, so long as you are having fun with friends.

Also, that fresh lemon to squeeze over, that was a new wrinkle to us for pasta, but we do it often now and love it. Try it with melted butter and then eat oatmeal for three days to clean out your arteries.

We were almost too contented to find our way back to St. Mark's Square. A gondola ride back to the bus/train station where the Orient Express sat waiting for an evening departure, topped off that lovely luncheon in a Venetian alley. Our romantic siestas lasted past dinner.

* * * * *

Probably the epitome of meals steeped in pure fun, the romance of strange cultures, and the experience of walking the very rooms where a famous person left an early legacy was a luncheon in Spain. The next tale and recipe should make your heart warm up with chuckles.

PICASSO, BEADY-EYED SHRIMP & SPANISH PULCHRITUDE

One of our band trips took us to the French Riviera and the delightfully historical town of Perpignan. We decided to dine in style one night, and hang the cost. The linen was crisp and snowy, the china collectible, the silver heavy and glowing. The waiter spoke no English and our limited French didn't quite cut the mustard. One of the women wanted veal chops in wine sauce, because she had had them once in Paris and loved them. But none of us knew how to ask for it, let alone read the menu.

At the risk of my limited dignity, I decided that the same pantomime I had used to obtain some suppositories for a suffering horn blower in Breganz, Austria might work here. In that delightful Austrian city, I had simply squatted down, grunted till I was red in the face and pointed to my fanny. The pharmacist immediately came up with a box of rectal suppositories. One of my dearest friends and the wife of the suffering band member for whom I was performing in public, pretended not to know me the rest of the day.

So, in this elegant Perpignan restaurant, I got the waiter's attention as I pointed to my rib cage, and the picture of a cow on the front of the menu (to let us foreigners know they served good beef), and then pretended to cradle a baby in my arms -- a baby cow, you know? Veal. My ribs, chops. The waiter smiled broadly, nodded, and hastened away. The rest of us managed to decipher the menu well enough to order our choices.

When our orders were served with a French Flourish we were all delighted. All, that is, except our veal chop lover. Veal chops in wine? Not by a long shot.

87

Sweetbreads, would you believe! Luckily, her husband adores the things and he happily traded his lamb and bean cassulet.

The next day, with my fame as an interpreter waning, we boarded the buses for a short trip over the Spanish border to the lovely little seaside village of Colliore, at the foot of the Pyrenees. We happened to choose a restaurant with a cobblestoned sidewalk cafe out front, facing a crowd of topless beauties and bikinied swains on the beach opposite. Our men were ecstatic! Most of them would have eaten anything their wives ordered. But not my city boy. I knew I would have to avoid anything with fish in it.

Being a tourist mecca, we had no language problem here. A sign even informed us in English that this restaurant's inside walls were lined with early Picasso paintings that had been traded by him for food and lodging back in his early starving artist days. Sadly, none were for sale!

It was here that I had my first and my very best Bouillabaisse ever, having been persuaded by our vocalist and lead sax man to share a bucket with him. My hubby's aversion to fish does not extend to boiled shrimp. So, I ordered a double platter for him.

When the bouillabaisse arrived, what a sight for hungry eyes! A shiny tin bucket, so hot even the bail was hot, and brim full of a spicy tomato broth with every kind of seafood you can imagine, and carrots and onions. **Note:** Regarding potatoes, see below.

Served alongside was a basket piled high with crispy chunks of French-type bread and a bowl of roasted garlic clusters fairly dripping with olive oil. We were shown how to squeeze the garlic over a chunk of bread, and dip it in the oil and then plop it into the soup. Doc and I ate the whole bucket, it must have been a gallon!

But, oh dear, when my hubby's shrimp arrived, they were piled willy-nilly on a platter with their feet and heads intact. Those beady eyes staring up at him sent my hero running off to the beach where the topless panorama helped to settle his tummy.

Our tenor sang all the way back to Perpignan while his darling wife wrote in her journal. I dozed against a sunny window, now and then catching some bawdy conversation from the back of the bus where my husband and the clarinet section shared their experiences collecting photos of the topless beach panorama.

My recipe is not as authentic as that bucket of glory, but we like it. **Note:** In Spain, they left the potatoes out, but served them on the side in a crockery bowl. We saw local folks spearing them and dunking them in the olive oil.

BOUILLABAISSE
(My laid back, easy way)

1 quart of tomato juice.
1 quart of Basic Tomato Sauce, or canned stewed tomatoes.
1 quart of chicken broth, defatted.
2 cups finely chopped onions.
1 cup finely chopped celery, leaves and all.
1/2 cup chopped parsley.
1 cup finely chopped green pepper.
1 cup firmly packed fresh basil leaves...fresh basil is imperrative...buy a couple plants and use it all.

About 1 dozen buds of garlic, crushed.
1/4 cup fresh chopped oregano, please use fresh.
About 1 dozen little bitzy red potatoes, well-
scrubbed.
5 or 6 fat carrots, cut in 1 inch chunks.
Salt and pepper to taste.
3/4 lb each, shelled shrimp, scallops, clams (I use
Progresso® canned, whole clams and use juice
too). If scallops and shrimp are huge, cut in half.
About 3/4 to 1 lb of any solid white fish, such as
turbot, dolphin, cobia, grouper. Cut the fish into 1
inch chunks.

If you like working at a meal and want to be utterly Spanish, you might add a bucket of mussels. But you will have to scrub their beards off, and then fish them out of the bowl with your fingers and dig them out of their shells. Forget it!

You will need your biggest soup, pasta, or canning pot. Start filling it in the order of the listed ingredients, up to the little red potatoes and carrots.

Bring slowly to a boil and reduce heat at once and simmer at least 45 minutes or until potatoes and carrots are tender. Then add seafood and simmer until the shrimp have turned a lovely bright pink. Stop cooking at once, or the shrimp will get tough. Salt and pepper to suit your tongue.

Have the bread basket loaded and a bowl of garlic roasted in oil and at the ready. Pass terrycloth tea towels for napkins. Serve a scoop of pineapple sherbet for dessert. If they are in season, sliced fresh peaches are wonderful too.

Now if there is any doubt your mind why our ancestors called tomatoes "Love Apples," you aren't reading me. This bright red berry which has been hybridized into all colors, and even a few new shapes, can become the basis of almost any flavor sensation you might be hankering after. And it has lasted through years and transcontinental miles to every table around which people who love people gather.

So gather up a couple of friends and go tomato picking, feel very good about yourself because you have saved money and done something productive and creative. I hope you will have stocked the fridge with gift items at the ready for a hostess, a care package for an elderly neighbor or that kid away from home. And what about your church or school bazaar? I have found that the savories and preserves sell well. Hitch your mind to fresh tomatoes and write your own book. You will have a blast, save money, and maybe even make spare change selling your goodies at fleamarkets.

Oh, save time for romance! Tomato flavored kisses are welcome beneath an olive tree in Spain, on a gondola in the Grand Canal in Venice, or in your own backyard in Ohio.

Index

A

C

D

E

F

G

THE ROMANTIC TOMATO

O

THE ROMANTIC TOMATO

The Romantic Tomato

How to Pick, Cook, Store & Enjoy Tomatoes

A practical book with mouthwatering tomato based recipes and humorous tales about traveling with THE AMERICAN WINDS CONCERT BAND; overnighting in Mom and Pop pensions outside major romantic mid-European cities, tasting our way in and out of these country kitchens.
LEARN -

* How to make the world's best, easiest basic sauces for ethnic dishes.
* How to use green tomatoes- FRIED.
* *GREEN TOMATO PIE.*
* *FAST, EASY KOSHER DILLED*, to slice.
* *DILLED KIRBY PICKLES TOO!*

If you hate tomatoes, buy the book anyway, who knows what'll happen next! (One gal won her soul mate with green tomato clam sauce.)

How To Board Up Your Kitchen And Cook From A Hammock

Brides, grads, retirees who are sick of cookin', laugh your way through pages of time and money saving tips. Easy parties, fast no work breakfasts, picnics, no cook, no bake desserts. You gotta have it!!

ORDER BLANK

No.	Item	Price	Total
	Books/Leaflets		
( )	*The Romantic Tomato**	$ 9.95	$_____
( )	*Board Up Your Kitchen**	$14.95	$_____
( )	Leaflets: Delicious new recipes for anti-cancer foods: carrots, cabbage, squash, kale, broccoli, etc. Beta carotene chart	$ 5.00	$_____
	Patterns/Instructions		
( )	TV Tray File System	$ 3.00	$_____
( )	Soft Storage	$ 3.00	$_____
( )	Easy Fabric Rugs	$ 3.00	$_____
		TOTAL	$_____

Florida residents must
add 6% state sales tax . . . . $_____

TOTAL ENCLOSED $_____

CHECK or MONEY ORDER payable to:
VIRGINIA B. ELLIOTT
P O Box 11983
Naples, FL 34101
Note: shipping is included in price.

BE SURE TO PRINT YOUR NAME, ADDRESS AND
NAMES OF PERSONS TO WHOM YOU WISH BOOKS TO
BE AUTOGRAPHED ON THE BACK OF THIS ORDER
BLANK.

YOUR NAME_____

STREET_____

CITY_____STATE____ZIP_____

NAMES FOR AUTOGRAPH & 10 WORD OR LESS MESSAGE.

1._____

_____

2._____

_____

3._____

_____

4._____

_____

5._____

_____

6._____

_____

7._____

_____

8._____

_____

* Order 3 or more BOOKS as gifts & get 10% discount on entire order

ORDER BLANK

No.	Item	Price	Total
	Books/Leaflets		
( )	*The Romantic Tomato**	$ 9.95	$_____
( )	*Board Up Your Kitchen**	$14.95	$_____
( )	Leaflets: Delicious new recipes for anti-cancer foods: carrots, cabbage, squash, kale, broccoli, etc. Beta carotene chart	$ 5.00	$_____
	Patterns/Instructions		
( )	TV Tray File System	$ 3.00	$_____
( )	Soft Storage	$ 3.00	$_____
( )	Easy Fabric Rugs	$ 3.00	$_____
		TOTAL	$_____

Florida residents must
add 6% state sales tax . . . . $_____

TOTAL ENCLOSED $_____

CHECK or MONEY ORDER payable to:
VIRGINIA B. ELLIOTT
P O Box 11983
Naples, FL 34101
Note: shipping is included in price.

BE SURE TO PRINT YOUR NAME, ADDRESS AND
NAMES OF PERSONS TO WHOM YOU WISH BOOKS TO
BE AUTOGRAPHED ON THE BACK OF THIS ORDER
BLANK.

YOUR NAME_____

STREET_____

CITY_____STATE____ZIP_____

NAMES FOR AUTOGRAPH & 10 WORD OR LESS MESSAGE.

1._____

_____

2._____

_____

3._____

_____

4._____

_____

5._____

_____

6._____

_____

7._____

_____

8._____

_____

* Order 3 or more <u>BOOKS</u> as gifts & get 10% discount on entire order